Collins

SNAP REVISION GEOMETRY AND MEASURES



# SNAP REVISION

## GEOMETRY AND MEASURES

(for papers 1, 2 and 3)

AQA GCSE Maths Foundation

AQA GCSE MATHS FOUNDATION

REVISE TRICKY TOPICS IN A SNAP

Maths

# Contents




Published by Collins
An imprint of HarperCollins*Publishers*
1 London Bridge Street,
London, SE1 9GF





9780008242374

First published 2017

10 9 8 7 6 5 4 3 2 1





British Library Cataloguing in Publication Data.

A CIP record of this book is available from the British Library.

Printed in Great Britain by Bell and Bain Ltd, Glasgow.

ACKNOWLEDGEMENTS

The author and publisher are grateful to the copyright holders for permission to use quoted materials and images.

All images are © Shutterstock.com

Every effort has been made to trace copyright holders and obtain their permission for the use of copyright material. The author and publisher will gladly receive information enabling them to rectify any error or omission in subsequent editions. All facts are correct at time of going to press.

# How To Use This Book

To get the most out of this revision guide, just work your way through the book in the order it is presented.

This is how it works:

**Revise** — Clear and concise revision notes help you get to grips with the topic

**Revise** — Key Points and Key Words explain the important information you need to know

**Revise** — A Quick Test at the end of every topic is a great way to check your understanding

**Practise** — Practice questions for each topic reinforce the revision content you have covered

**Review** — The Review section is a chance to revisit the topic to improve your recall in the exam

# Angles and Shapes 1

**You must be able to:**

- Recognise relationships between angles
- Use the properties of angles to work out unknown angles
- Recognise different types of triangle
- Understand and use the properties of special types of quadrilaterals.

## Angle Facts

- There are three types of angle:
  - **acute**: less than 90°
  - **obtuse**: between 90° and 180°
  - **reflex**: between 180° and 360°.
- Angles on a straight line add up to 180°.
- Angles around a point add to 360°.
- **Vertically opposite** angles are equal.

## Angles in Parallel Lines

- Parallel lines never meet. The lines are always the same distance apart.
- **Alternate** angles are equal.
- **Corresponding** angles are equal.
- Co-interior or **allied** angles add up to 180°.

Work out the sizes of angles $a$, $b$, $c$ and $d$.
Give reasons for your answers.

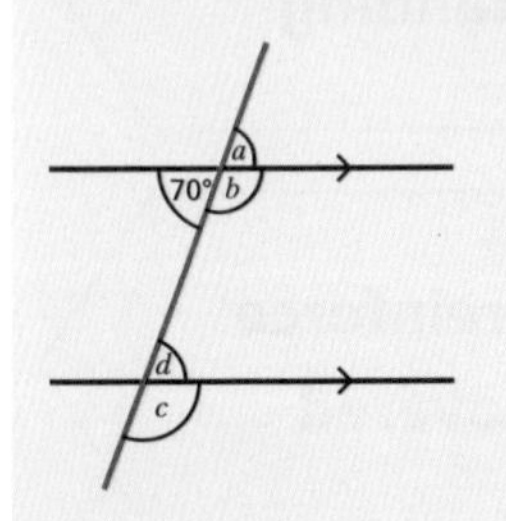


$a = 70°$ (vertically opposite angles are equal)

$b = 110°$ (angles on a straight line add up to 180°, so $b = 180° - 70°$)

$c = 110°$ (corresponding to $b$; corresponding angles are equal)

$d = 70°$ (corresponding to $a$; corresponding angles are equal)

## Triangles

- Angles in a triangle add up to 180°.
- There are several types of triangle:
  - **equilateral**: three equal sides and three equal angles of 60°
  - **isosceles**: two equal sides and two equal angles (opposite the equal sides)
  - **scalene**: no sides or angles are equal
  - **right-angled**: one 90° angle.

**Alternate Angles**

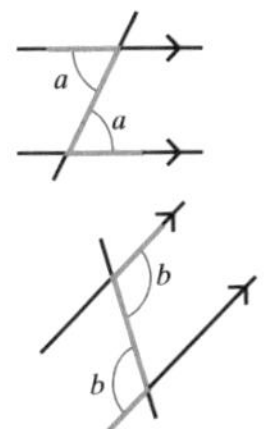


**Corresponding Angles**

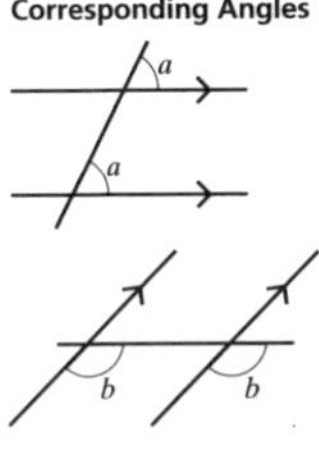


**Allied Angles**

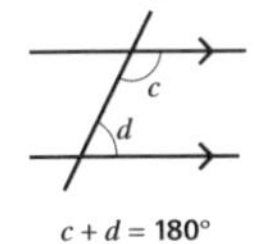


**Key Point**

Examiners will **not** accept terms like 'Z angles' or 'F angles'. Always use correct terminology when giving reasons.

*ABC* is an isosceles triangle and *HE* is parallel to *GD*.
*BAF* is a straight line. Angle *FAE* = 81°

Calculate **a)** angle *ABC* and **b)** angle *ACB*.
Give reasons for your answers.

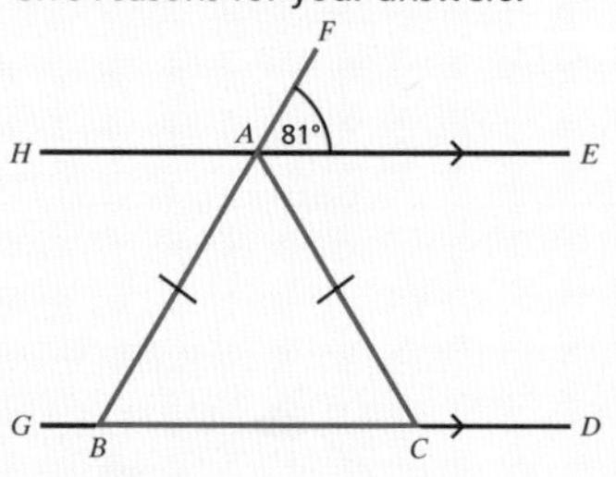


**a)** Angle *HAB* = 81° (vertically opposite *FAE*), so angle *ABC* = 81° (alternate angle to *HAB*)

**b)** Angle *ACB* = 81° (angle *ABC* = angle *ACB*; base angles of an isosceles triangle are equal.)

There are several different ways of solving this question.

## Special Quadrilaterals

- The interior angles in a quadrilateral add up to 360°.
- The order of rotational symmetry is the number of times a shape looks the same when it is rotated 360° (one full turn).
- You need to know the properties of these special quadrilaterals:

| | Sides | Angles | Lines of Symmetry | Rotational Symmetry | Diagonals |
|---|---|---|---|---|---|
| **parallelogram** | opposite sides are equal and parallel | diagonally opposite angles are equal | none | order 2 | diagonals bisect each other |
| **rhombus** | all sides are equal and opposite sides are parallel | opposite angles are equal | two | order 2 | diagonals bisect each other at 90° |
| **kite** | two pairs of adjacent sides are equal | one pair of opposite angles are equal | one | none | diagonals cross at 90° |
| **trapezium** | one pair of opposite sides are parallel | | none (an isosceles trapezium has one) | none | |

### Quick Test

1. Name all the quadrilaterals that can be drawn with lines of lengths:
   a) 4cm, 7cm, 4cm, 7cm  b) 6cm, 6cm, 6cm, 6cm
2. *EFGH* is a trapezium with *EH* parallel to *FG*.
   *FE* and *GH* are produced (made longer) to meet at *J*.
   Angle *EHF* = 62°, angle *EFH* = 25° and angle *JGF* = 77°.
   Calculate the size of angle *EJH*.

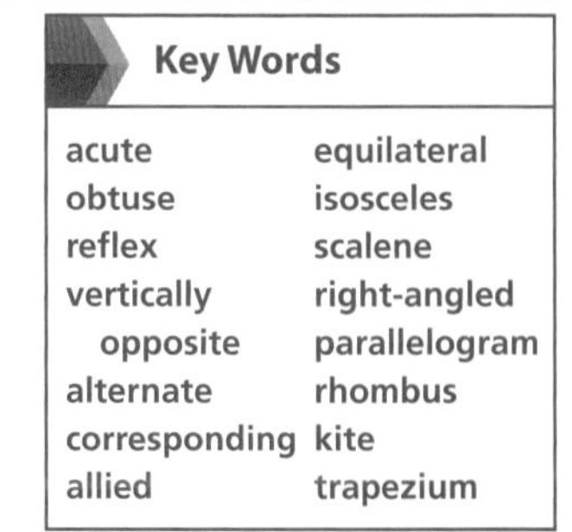

### Key Words

acute
obtuse
reflex
vertically opposite
alternate
corresponding
allied
equilateral
isosceles
scalene
right-angled
parallelogram
rhombus
kite
trapezium

# Angles and Shapes 2

**You must be able to:**

- Work out angles in a polygon
- Answer questions on regular polygons
- Understand scale drawings and use bearings.

## Angles in a Polygon

- A **polygon** is a closed shape with at least three straight sides.
- **Regular** polygons are shapes where all the sides and angles are equal.
- **Irregular** polygons are shapes where some or all of the sides and angles are different.
- For all polygons:
  - at any **vertex** (corner): **interior** angle + **exterior** angle = 180°
  - sum of all exterior angles = 360°.
- To work out the sum of the interior angles in a polygon, you can split it into triangles from one vertex.
- For example, a pentagon can be divided into three triangles, so the sum of the interior angles is $3 \times 180° = 540°$.
- The sum of the interior angles for any polygon can be calculated using the formula:

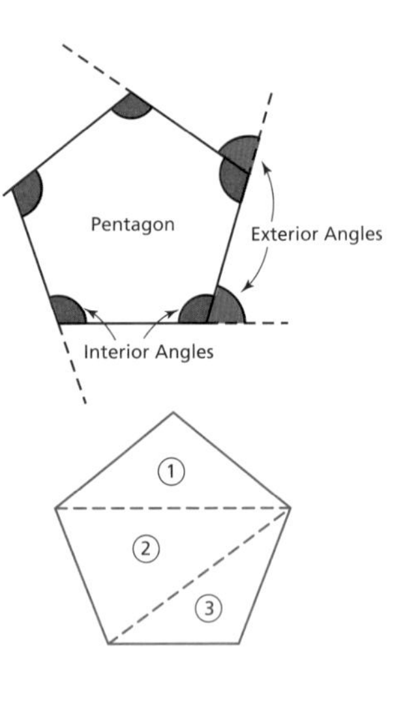

LEARN

Sum = $(n - 2) \times 180°$  Where $n$ = number of sides

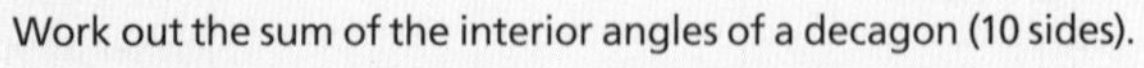

Work out the sum of the interior angles of a decagon (10 sides).

Sum = $(10 - 2) \times 180°$
= $8 \times 180°$
= 1440°

Use the formula:
Sum = $(n - 2) \times 180°$

## Regular Polygons

- In regular polygons:

LEARN

Number of Sides ($n$) × Exterior Angle = 360°
So, Exterior Angle = $360° \div n$

Work out the size of the interior angles in a regular hexagon (six sides).

Exterior angle = $360° \div 6 = 60°$
Interior angle + 60° = 180°
Interior angle = 180° − 60°
= 120°

Use the formula:
Exterior Angle = $360° \div n$

Interior Angle + Exterior Angle = 180°

A regular polygon has an interior angle of 156°.

Work out the number of sides that the polygon has.

Exterior angle = 180° – interior angle
= 180° – 156° = 24°

Number of sides = 360° ÷ 24° = 15

## Scale Drawings and Bearings

- **Bearings** are always measured in a clockwise direction from north (000°) and have three figures.

This is a radar screen showing the position of four aircraft.
Scale 1mm : 4km

1mm on the diagram represents 4km in real life.

Describe the locations of aircraft A, B, C and D in relation to the centre (airport).

Aircraft A is on a bearing of 030° and 80km from the airport.

Aircraft B is on a bearing of 210° and 60km from the airport.

Aircraft C is on a bearing of 330° and 100km from the airport.

Aircraft D is on a bearing of 120° and 40km from the airport.

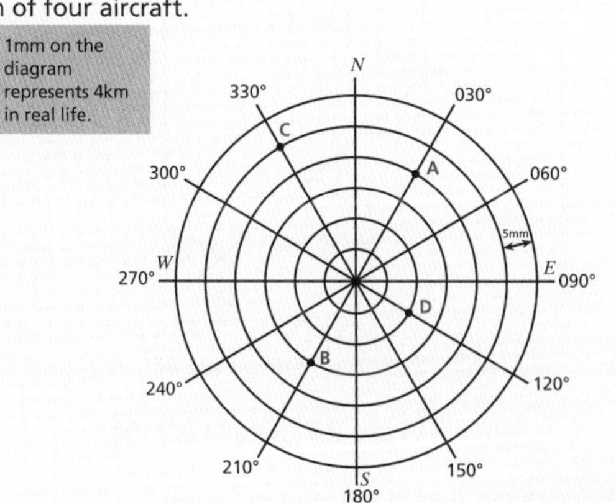


A ship sails from Mevagissey on a bearing of 130° for 22km.

**a)** Draw an accurate diagram to show this information and state the scale you have used.

**b)** What bearing would take the ship back to the harbour?

New bearing to return to harbour = 310°

Measure with a protractor.

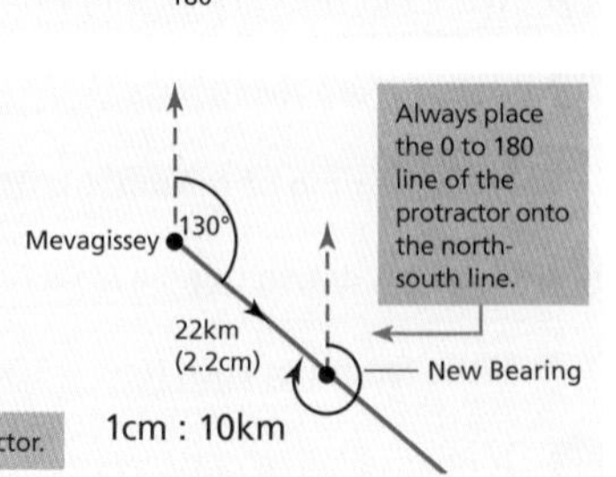

### Quick Test

1. For a regular icosagon (20 sides), work out **a)** the sum of the interior angles and **b)** the size of one interior angle.
2. A regular polygon has an interior angle of 150°. How many sides does the polygon have?
3. Two yachts leave port at the same time. Yacht A sails on a bearing of 040° for 35km. Yacht B sails on a bearing of 120° for 60km. Using a scale of 1cm : 10km, draw the route taken by both yachts. Measure the bearing of yacht B from yacht A.

### Key Words

**polygon**
**regular**
**irregular**
**vertex**
**interior**
**exterior**
**bearing**

# Practice Questions

## Angles and Shapes 1 & 2

1. Work out the size of angles $j$, $k$, $l$ and $m$, giving a reason for each answer. 

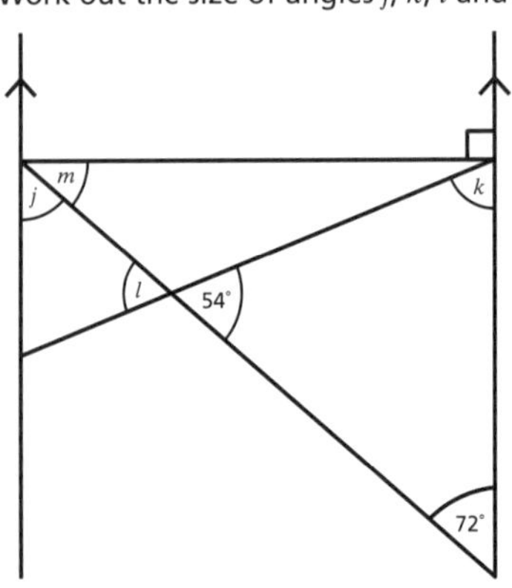


[4]

2. Three angles in a quadrilateral are 46°, 107° and 119°.

   Calculate the size of the fourth angle. [1]

3. Work out the exterior angle of a regular decagon. [1]

4. $A$ and $B$ are two points.

   If the bearing of $B$ from $A$ is 036°, what is the bearing of $A$ from $B$? [1]

5. A map is drawn using a scale of 1cm : 4km.

   If the length of Loch Ness is 36km, what would its length be on the map? [1]

6. A boat sails in a north-westerly direction.

   What bearing is this? [1]

**Total Marks** ........................ / 9

## Angles and Shapes 1 & 2

**1** Work out the value of $x$.

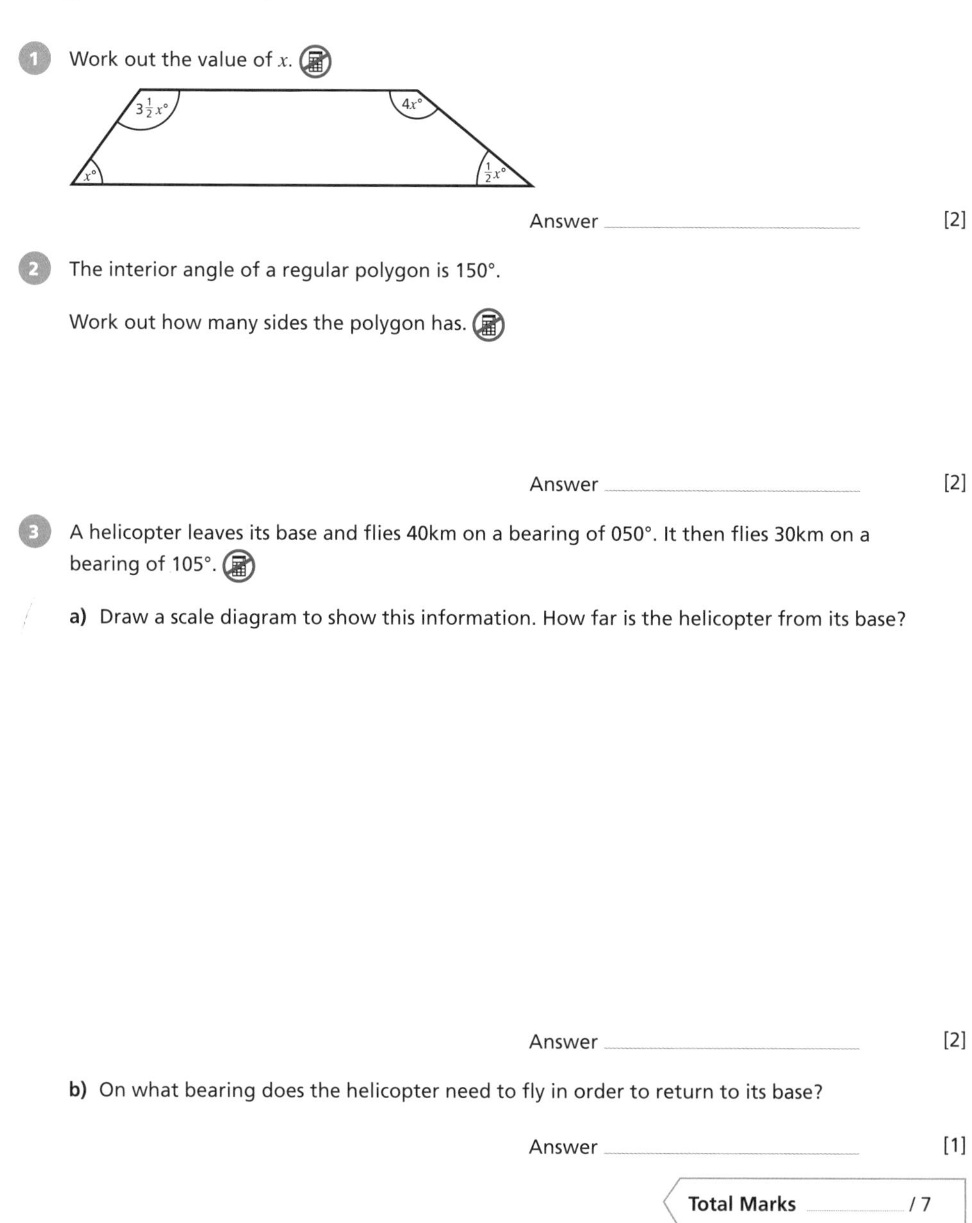


Answer ............................................ [2]

**2** The interior angle of a regular polygon is 150°.

Work out how many sides the polygon has.

Answer ............................................ [2]

**3** A helicopter leaves its base and flies 40km on a bearing of 050°. It then flies 30km on a bearing of 105°.

**a)** Draw a scale diagram to show this information. How far is the helicopter from its base?

Answer ............................................ [2]

**b)** On what bearing does the helicopter need to fly in order to return to its base?

Answer ............................................ [1]

**Total Marks** .................... / 7

# Transformations

**You must be able to:**

- Identify, describe and construct transformations of shapes, including reflections, rotations, translations and enlargements.

## Reflection

- When a shape is reflected:
  - Each point on the image is the same distance from the mirror line as the corresponding point on the object
  - The object and the image are **congruent** (same size and shape)
  - To define a **reflection** on a coordinate grid, the equation of the mirror line should be stated.

## Rotation

- **Rotation** is described by stating the:
  - Direction rotated (clockwise or anticlockwise)
  - Angle of rotation (which is usually a multiple of 90° in the exam)
  - Centre of rotation (point about which the shape is rotated).

**Key Point**

There is no need to state clockwise or anticlockwise for a rotation of 180°.

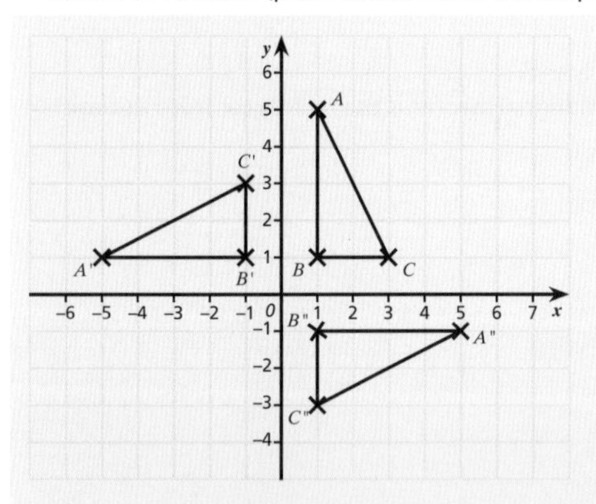


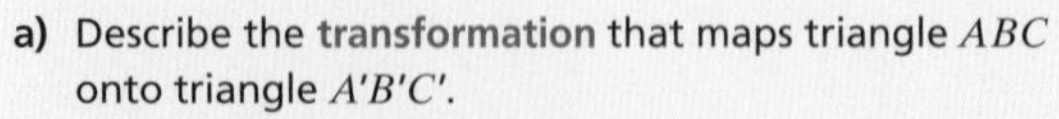

**a)** Describe the **transformation** that maps triangle $ABC$ onto triangle $A'B'C'$.

The transformation that maps $ABC$ to $A'B'C'$ is a rotation 90° anticlockwise (or 270° clockwise) about the origin (0, 0).

**b)** Describe the transformation that maps triangle $A'B'C'$ onto triangle $A''B''C''$.

The transformation that maps $A'B'C'$ to $A''B''C''$ is a rotation of 180° about the origin (0, 0).

# Translation

- When a shape is translated:
  - The shape does not rotate – it moves left or right and up or down – and stays the same size
  - The **translation** is represented by a **column vector** $\begin{pmatrix} x \\ y \end{pmatrix}$.

Describe the transformation that takes shape A to shape B.

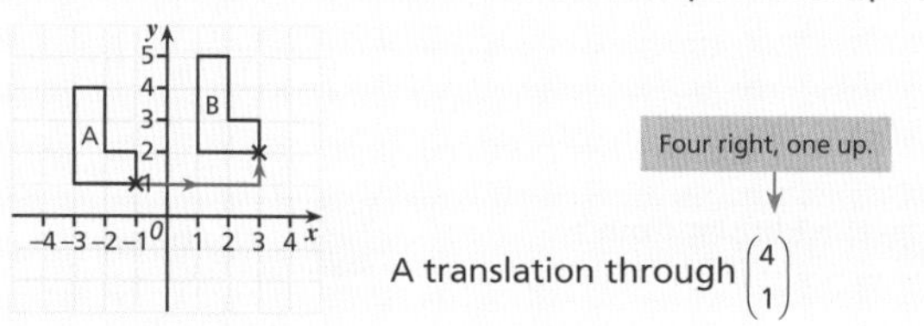


A translation through $\begin{pmatrix} 4 \\ 1 \end{pmatrix}$

**Key Point**

$x$ represents the distance moved **horizontally: positive** means to the **right** and **negative** means to the **left**.

$y$ represents the distance moved **vertically: positive** means **up** and **negative** means **down.**

# Enlargement

- When a shape is enlarged:
  - The shape of the object is not changed, only its size
  - The enlarged shape is **similar** to the original shape
  - The **scale factor** determines whether the object gets bigger (scale factor > 1) or smaller (scale factor < 1)
  - When describing the **enlargement**, state the scale factor and the centre of enlargement.

**Key Point**

Scale factors that result in the object getting smaller are often represented as a fraction.

**a)** Enlarge triangle A by scale factor 2, centre of enlargement (1, 2). Label the transformed triangle, B.

**b)** Enlarge triangle A by scale factor $\frac{1}{2}$, centre of enlargement (1, 2). Label the transformed triangle, C.

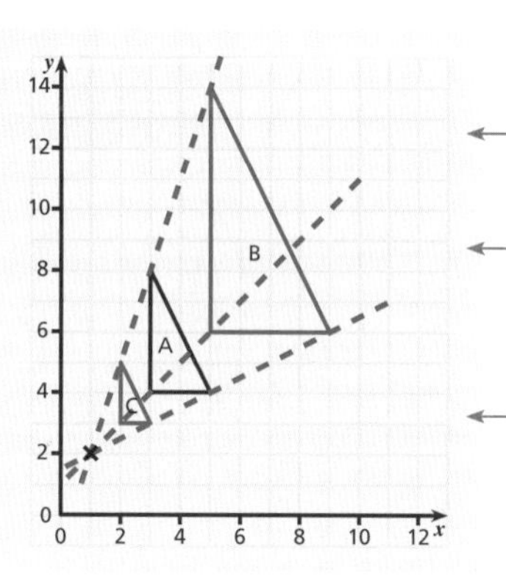


All construction lines must remain.

The side **lengths** of triangle B are **twice** the length of the corresponding sides of triangle A. However, the **area** of triangle B is **four times** bigger.

The side **lengths** of triangle C are **half** the length of the corresponding sides of triangle A. However, the **area** of triangle C is **four times** smaller.

**Key Words**

congruent
reflection
rotation
transformation
translation
column vector
similar
scale factor
enlargement

**Quick Test**

1. Describe the single transformation that takes:
   a) Triangle A to triangle B
   b) Triangle A to triangle C
   c) Triangle A to triangle D.

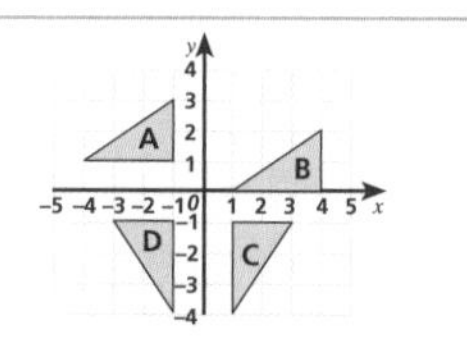

# Constructions

**You must be able to:**

- Use a ruler and a pair of compasses to produce different constructions, including bisectors
- Describe a locus and solve problems involving loci.

## Constructions

- The **perpendicular bisector** of line $AB$.

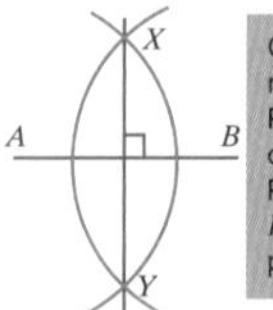


Open compasses to more than half $AB$. Put compass point on $A$. Draw arc. Put compass point on $B$. Draw arc. $XY$ is the perpendicular bisector.

- A **perpendicular from a given point** to the line $AB$.

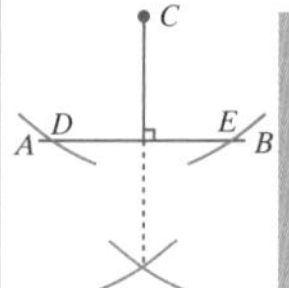


Put compass point on $C$. Draw arc. Keep radius the same. Put compass point on $E$. Draw arc. Put compass point on $D$. Draw arc. Join $C$ to the point where the arcs cross.

- An **angle bisector.**

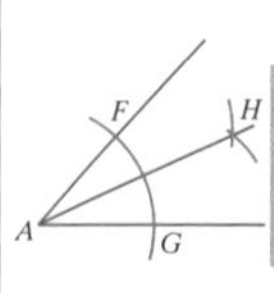


Put compass point on $A$. Draw arc $FG$. Put compass point on $F$. Draw arc. Put compass point on $G$. Draw arc. Join $HA$.

- An **equilateral triangle** and an **angle of 60°.**

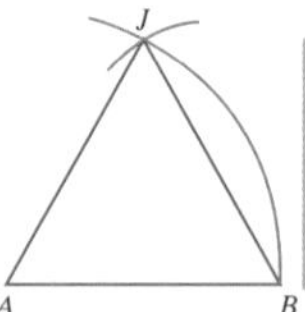


Open compasses to length $AB$. Put point on $A$. Draw arc. Put compass point on $B$. Draw arc. Join $AJ$ and $BJ$. Angle $A = 60°$

- A **perpendicular** to line $AB$ **at a given point**.

Put compass point on $M$. Draw arcs at $K$ and $L$. Construct the perpendicular bisector of $K$ and $L$.

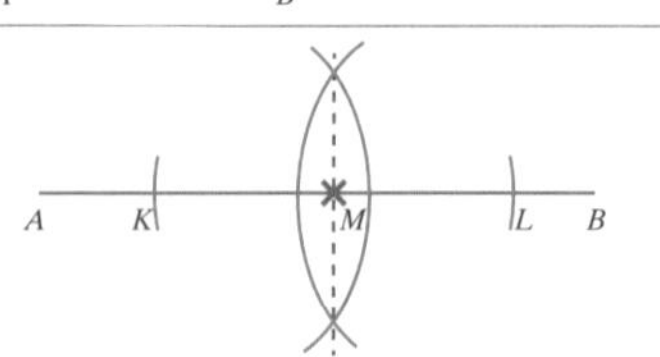


### Key Point

The perpendicular distance from a point to a line is the shortest distance to the line.

## Defining a Locus

- A **locus** is the path taken by a point that is obeying certain rules.
- The plural of locus is **loci**.

- The locus of points that are a **fixed distance from a given point,** $A$, is a circle.

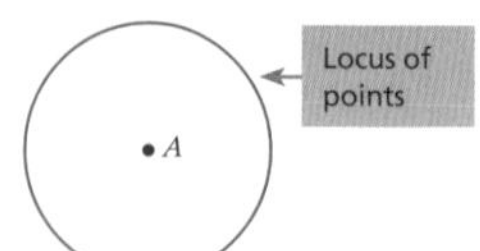


Locus of points

- The locus of points that are a **fixed distance from a line** $AB$.

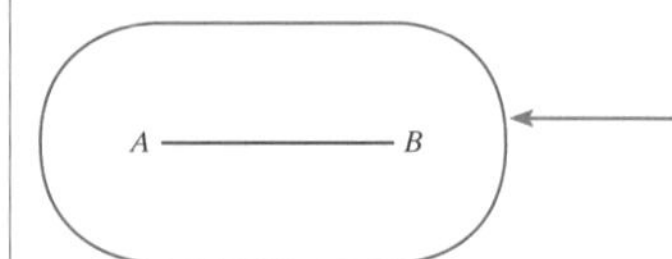


Locus of points

### Key Point

In the exam, the fixed distance is likely to be given, e.g. 4cm.

- The locus of points that are the **same distance from two lines** $AB$ and $AC$. This is the angle bisector.

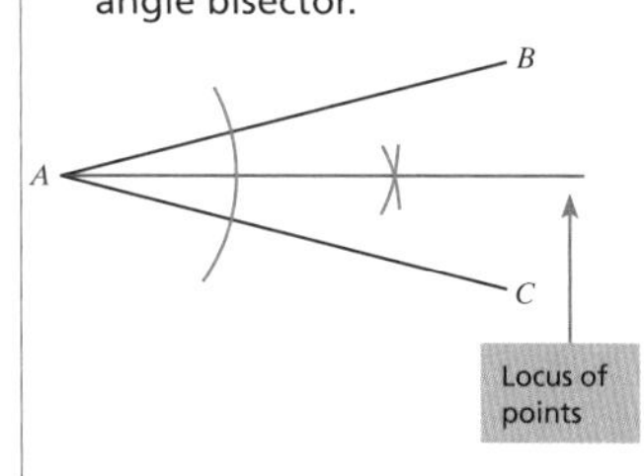


- The locus of points that are **equidistant from two points** $A$ and $B$. This is the perpendicular bisector.

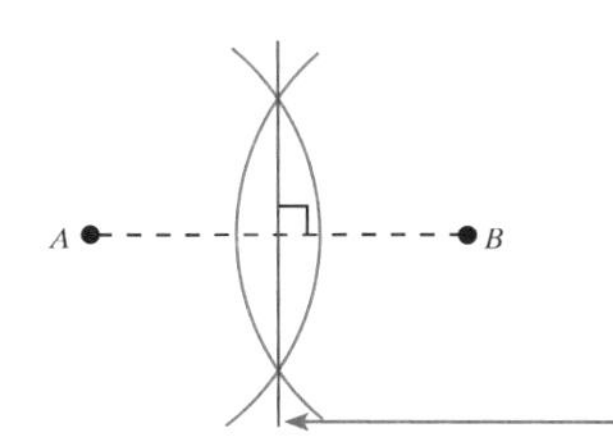


Locus of points

**Key Point**

Equidistant means 'of equal distance'.

## Loci Problems

A guard dog is tied to a post by a 4-metre long rope.

Accurately draw the locus of the points the dog can reach using a scale of 1cm : 1m.

The solution would be a shaded circle of radius 4cm.
The dog could reach the circumference of the circle and any point inside it.

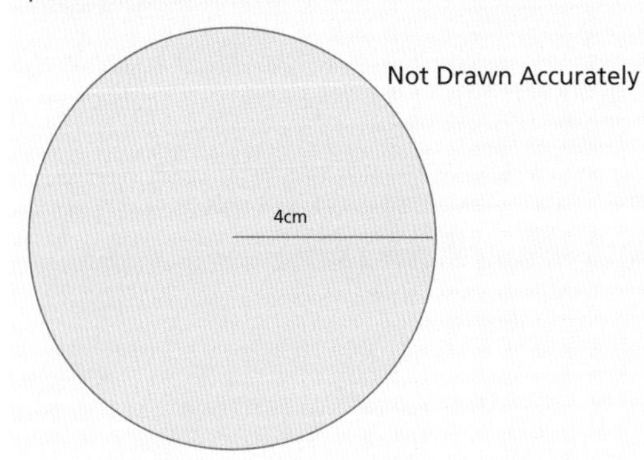


**Quick Test**

1. Describe how you would accurately construct an angle of 30°.
2. A rectangular vegetable plot $ABCD$ measures 6m by 3m. A goat is tethered at corner $A$ by a 4m rope. Accurately draw the plot and construct the locus of the points that the goat can reach. Scale 1cm : 1m. Shade the region of the vegetable plot that can be eaten by the goat.

**Key Words**

**perpendicular bisector**
**locus / loci**
**equidistant**

# Nets, Plans and Elevations

**You must be able to:**

- Identify a 3D shape from its net
- Draw nets of 3D shapes
- Use isometric grids
- Interpret and draw plans and elevations of 3D shapes.

## Nets

- A **net** is a 2D shape that can be folded to form a 3D solid.

| Name | Shape | Net |
|---|---|---|
| Cube | 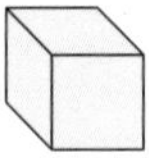 | 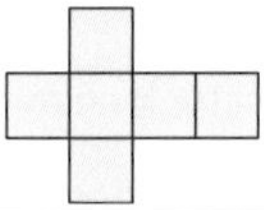 |
| Cuboid | 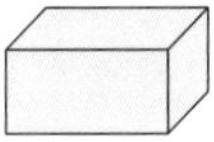 | 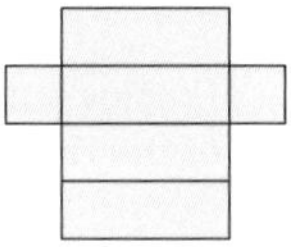 |
| Cylinder | 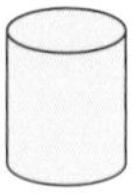 | 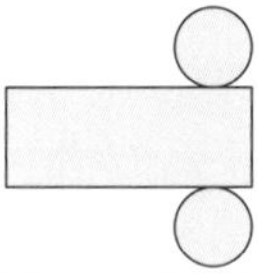 |
| Triangular Prism | 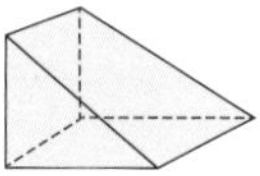 | 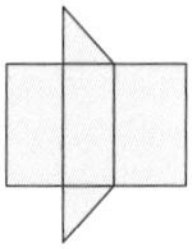 |
| Square-Based Pyramid | 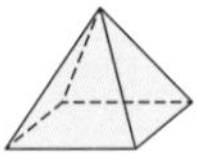 | 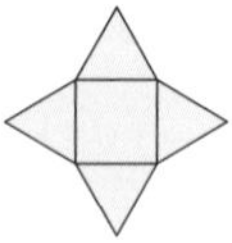 |

### Key Point

There are many different nets for a cube.

## Plans and Elevations

- An **isometric grid** can be used to draw 3D shapes.
- The **plan view** of a 3D shape shows what it looks like from above, i.e. a bird's eye view.
- The side **elevation** is the view of a 3D shape from the side.
- The front elevation is the view of a 3D shape from the front.

### Key Point

An isometric grid can be made up of equilateral triangles or dots.

Here is a 3D shape made of centimetre cubes:

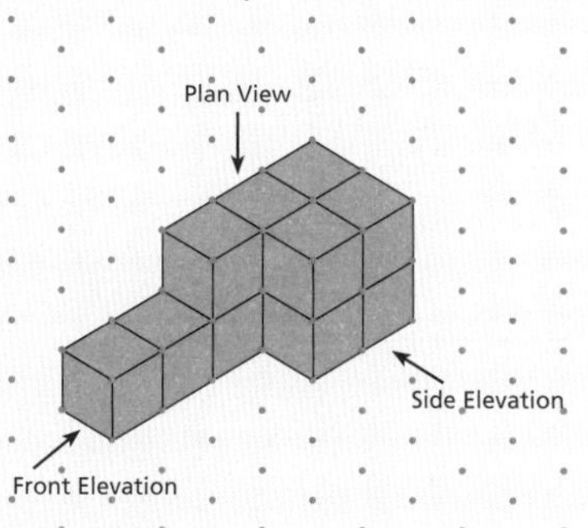


Draw:

a) The plan view

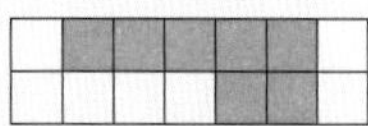

Count the number of cubes for each length carefully.

b) The front elevation

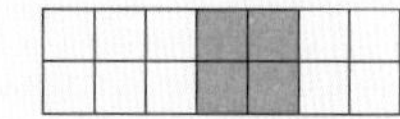

c) The side elevation.

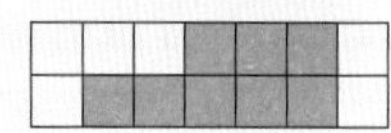

Here are the three views of a 3D shape:

Plan View

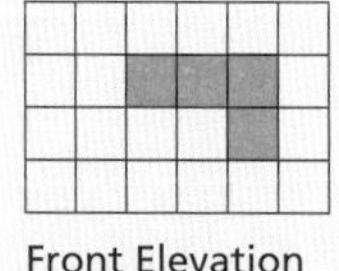

Front Elevation

Side Elevation

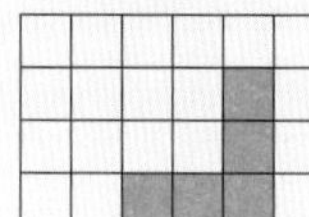

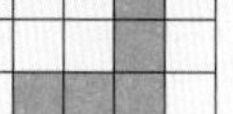

Draw the 3D shape on isometric paper.

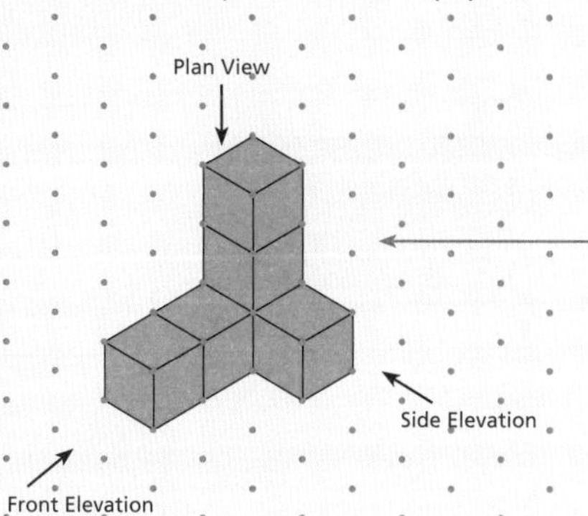


Check that your final drawing matches the given views.

## Quick Test

1. A can of soup has a height of 10cm. The length of the label required to go once around the can is 22cm.

Draw an accurate plan view and front elevation of the can of soup.

## Key Words

net
isometric grid
plan view
elevation

# Area and Volume 1

**You must be able to:**

- Calculate the perimeter and area of rectangles and triangles
- Calculate the perimeter and area of composite shapes
- Calculate the volume and surface area of a cuboid.

## Rectangles

- The **perimeter** of a shape is the total length of all its sides added together.
- The **area** is the space enclosed within the perimeter.

**LEARN**

Area of a Rectangle ($A$) = Length ($l$) × Width ($w$)

$$A = lw$$

Perimeter of a Rectangle ($P$) = 2 × Length ($l$) + 2 × Width ($w$)

$$P = 2l + 2w$$

Calculate the perimeter and area of the rectangle.

5cm

8cm

**Area**

$A = 5 \times 8$
$= 40\text{cm}^2$

**Perimeter**

$P = 2 \times 5 + 2 \times 8$
$= 10 + 16 = 26\text{cm}$

## Triangles

**LEARN**

Area of a Triangle ($A$) = $\frac{1}{2}$ × Base ($b$) × Height ($h$)

$$A = \frac{1}{2}bh$$

Work out the perimeter and area of the triangle.

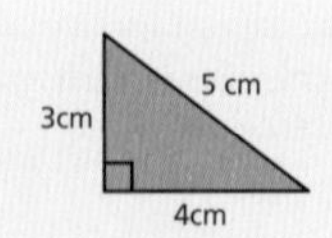


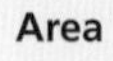

**Area**

$A = \frac{1}{2}bh = \frac{1}{2} \times 4 \times 3$
$= 6\text{cm}^2$

**Perimeter**

$P = 3 + 4 + 5$
$= 12\text{cm}$

**Key Point**

When calculating the area of a triangle, always use the perpendicular height.

## Composite Shapes

- **Composite shapes** are made up of other shapes.
- To find the perimeter and area of composite shapes, break them down into their component shapes.

Calculate the perimeter and area of the shape below.

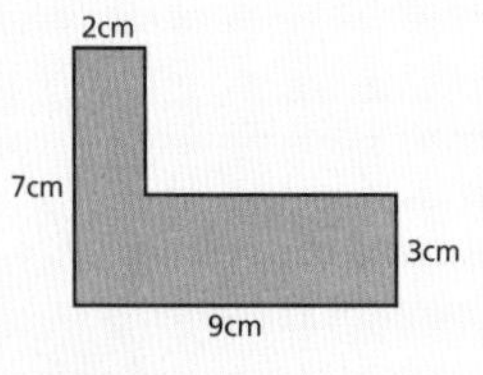


**Area**

$A = 7 \times 2 + 7 \times 3$

$= 14 + 21$

$= 35\text{cm}^2$

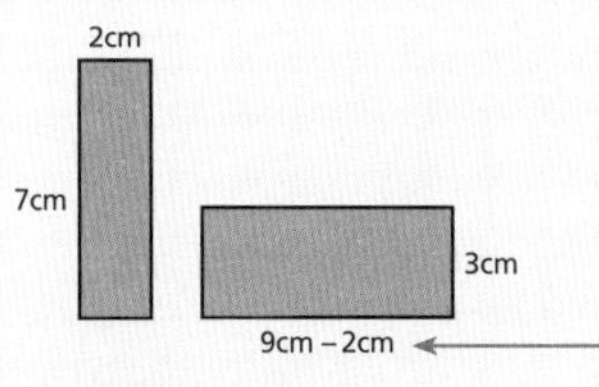


Work out the unknown length.

**Perimeter**

$P = 2 + 7 + 9 + 3 + 7 + 4$

$= 32\text{cm}$

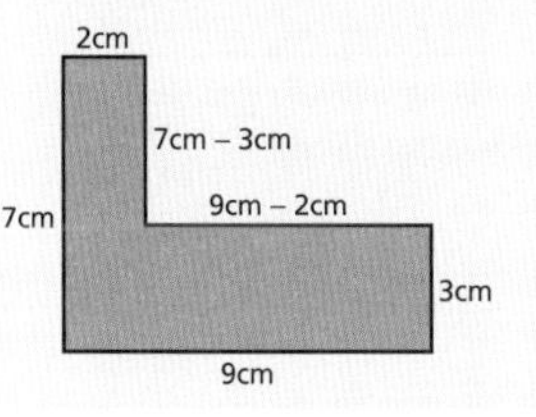


> **Key Point**
>
> When calculating area, remember to give your answer in units², e.g. cm².

## Cuboids

Volume of a Cuboid ($V$) = Length ($l$) × Width ($w$) × Height ($h$)

$$V = lwh$$

Surface Area of a Cuboid ($SA$) = $2lw + 2lh + 2hw$

> **Key Point**
>
> When calculating volume, remember to give your answer in units³, e.g. cm³.

Work out the volume and surface area of the cuboid.

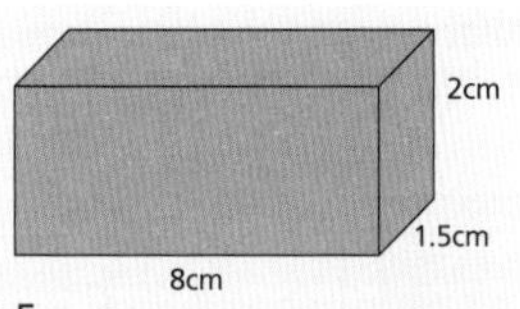
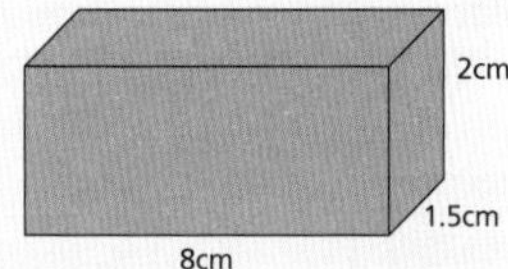


$V = 8 \times 1.5 \times 2$

$= 24\text{cm}^3$

$SA = 2 \times 8 \times 1.5 + 2 \times 8 \times 2 + 2 \times 2 \times 1.5$

$= 24 + 32 + 6$

$= 62\text{cm}^2$

**Quick Test**

1. A rectangle has a length of 4m and a width of 3m. Calculate the area and perimeter.
2. A triangle has a base of length 5cm and a perpendicular height of 4cm. Calculate the area.
3. A cuboid has a length of 3cm, a width of 3cm and a height of 2cm. Work out the volume and surface area of the cuboid.

**Key Words**

**perimeter**
**area**
**composite shapes**

# Area and Volume 2

**You must be able to:**

- Recall and use the formulae for the circumference and area of a circle
- Recall and use the formula for the area of a trapezium
- Recall and use the formulae for the volume and surface area of a prism
- Recall and use the formulae for the volume and surface area of a cylinder.

## Circles

LEARN

Circumference of a Circle ($C$) = $2\pi r$ or $C = \pi d$

Area of a Circle ($A$) = $\pi r^2$

Work out the circumference and area of a circle with radius 9cm. Give your answers to 1 decimal place.

**Circumference**

$C = 2 \times \pi \times 9$

$= 18 \times \pi$

$= 56.5$cm (to 1 d.p.)

**Area**

$A = \pi \times 9^2$

$= \pi \times 81$

$= 254.5\text{cm}^2$ (to 1 d.p.)

**Key Point**

The symbol $\pi$ represents the number **pi**.

$\pi$ can be approximated by 3.14 or $\frac{22}{7}$.

## Trapeziums

LEARN

The area of a **trapezium** is:

$A = \frac{1}{2}(a + b)h$

where $a$ and $b$ are the **parallel** sides and $h$ is the **perpendicular** height

- This formula can be proved:

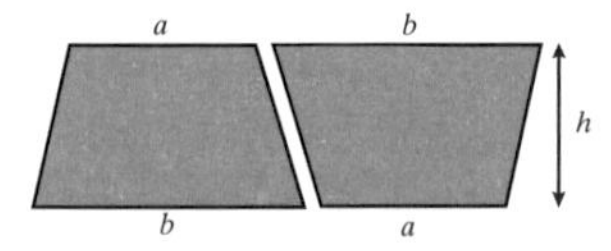


  - Two identical trapeziums fit together to make a parallelogram with base $a + b$ and height $h$
  - The area of the parallelogram is $(a + b)h$
  - Therefore, the area of each trapezium is $\frac{1}{2}(a + b)h$.

Work out the area of the trapezium.

$A = \frac{1}{2} \times (5 + 10) \times 4$

$= 30\text{cm}^2$

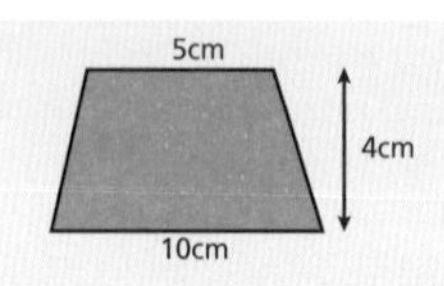


**Key Point**

Perpendicular means 'at right angles'.

Parallel means 'in the same direction and always the same distance apart'.

**Key Point**

The area of a parallelogram is: $A = bh$

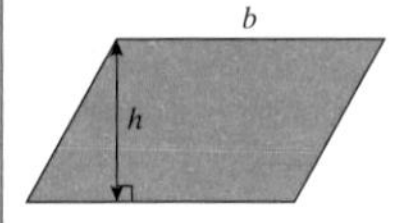

## Prisms

- A right prism is a 3D shape that has the same **cross-section** running all the way through it.

Volume of a Prism = Area of Cross-Section × Length

- The surface area is the sum of the areas of all the **faces**.

Work out the volume and surface area of the triangular prism.

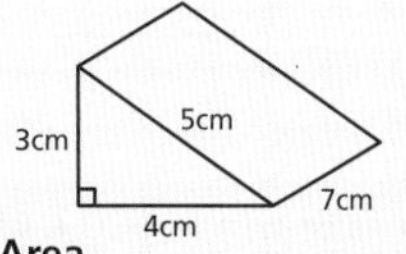


**Volume**

Area of the cross-section $= \frac{1}{2} \times 3 \times 4 = 6\text{cm}^2$

Volume $= 6 \times 7$
$= 42\text{cm}^3$

**Surface Area**

Five faces:

Two triangular faces = 6 + 6 = 12
Base = 4 × 7 = 28
Side = 3 × 7 = 21
Slanted side = 5 × 7 = 35

Total surface area =
12 + 28 + 21 + 35 = $96\text{cm}^2$

## Cylinders

Volume of a Cylinder = $\pi r^2 h$

Surface Area of a Cylinder = $2\pi rh + 2\pi r^2$

Work out the volume and the surface area of the cylinder. Give your answers in terms of $\pi$.

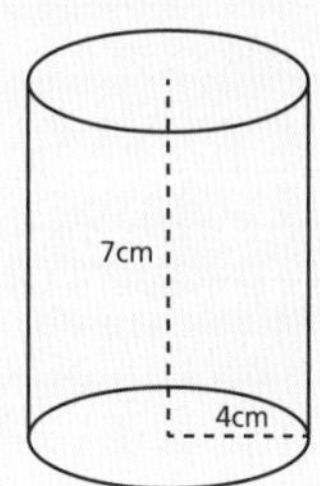


**Volume**

$V = \pi \times 4^2 \times 7$
$= 112\pi\text{cm}^3$

**Surface Area**

$SA = 2 \times \pi \times 4 \times 7 + 2 \times \pi \times 4^2$
$= 56\pi + 32\pi$
$= 88\pi\text{cm}^2$

### Key Point

A cylinder is just like any other right prism. To find the volume, you multiply the area of the cross-section (circular face) by the length of the cylinder.

### Quick Test

1. Calculate the volume and surface area of a cylinder with radius 4cm and height 6cm.
2. Work out the area of the trapezium.

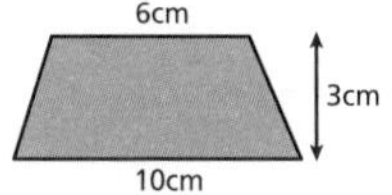


3. Calculate the circumference and area of a circle, diameter 7cm.

### Key Words

**trapezium**
**parallel**
**perpendicular**
**cross-section**
**face**

# Area and Volume 3

- **You must be able to:**
- Find the volume of a pyramid
- Find the volume and surface area of a cone
- Find the volume of a frustum
- Find the volume and surface area of a sphere
- Find the area and volume of composite shapes.

## Pyramids

- A **pyramid** is a 3D shape in which lines drawn from the **vertices** of the base meet at a point.

Volume of a Pyramid = $\frac{1}{3}$ × Area of the Base × Height

**Key Point**

A pyramid is usually defined by the base, e.g. a square-based pyramid or a triangular-based pyramid.

Work out the volume of the square-based pyramid.

$V = \frac{1}{3} \times 9 \times 9 \times 7$

$= 189\text{cm}^3$

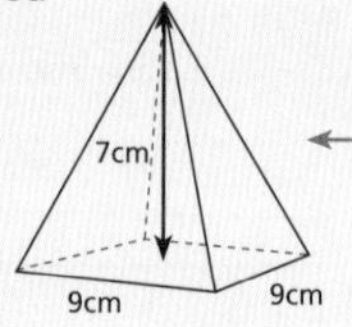


You must use the perpendicular height to calculate volume.

## Cones

- A **cone** is a 3D shape with a circular base that tapers to a single vertex.

Volume of a Cone = $\frac{1}{3}\pi r^2 h$

Surface Area of a Cone = $\pi rl + \pi r^2$

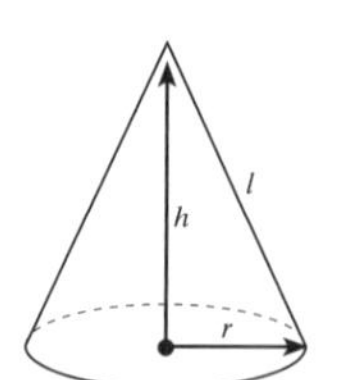


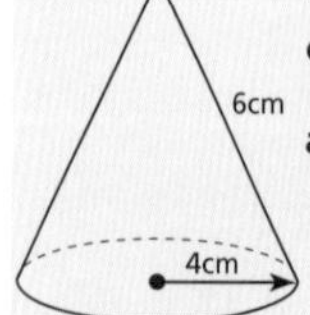
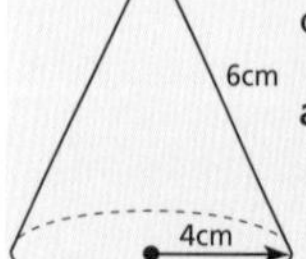


Work out **a)** the volume and **b)** the surface area of the cone. Give your answers to 1 decimal place.

**a)** $h = \sqrt{6^2 - 4^2}$

$h = \sqrt{20}$

$V = \frac{1}{3} \times \pi \times 4^2 \times \sqrt{20} = 74.9\text{cm}^3$

**b)** $SA = (\pi \times 4 \times 6) + (\pi \times 4^2) = 125.7\text{cm}^2$

First find the height using Pythagoras' Theorem (see p.28).

- A **frustum** is the 3D shape that remains when a cone is cut parallel to its base and the top cone removed.
- The original cone and the smaller cone that is removed are always **similar**.

Volume of a Frustum
= Volume of Whole Cone – Volume of Top Cone

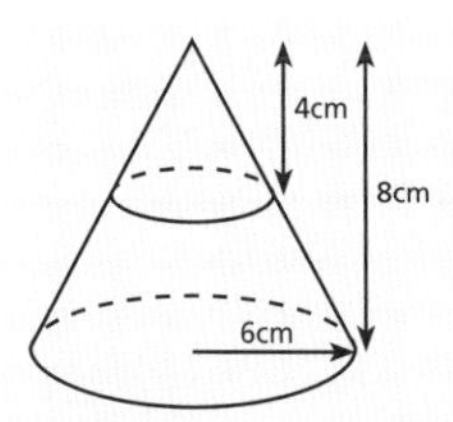


Calculate the volume of the frustum. Leave your answer in terms of $\pi$.

Radius of small cone = 3cm

$V = \frac{1}{3}(\pi \times 6^2 \times 8) - \frac{1}{3}(\pi \times 3^2 \times 4)$

$= 84\pi \text{cm}^3$

The two cones are similar with scale factor 2.

## Spheres

- A **sphere** is a 3D shape that is round, like a ball. At every point, its surface is equidistant from its centre.

Volume of a Sphere = $\frac{4}{3}\pi r^3$

Surface Area of a Sphere = $4\pi r^2$

- A **hemisphere** is half of a sphere; a dome with a circular base.

Work out **a)** the volume and **b)** the surface area of the sphere. Leave your answers in terms of $\pi$.

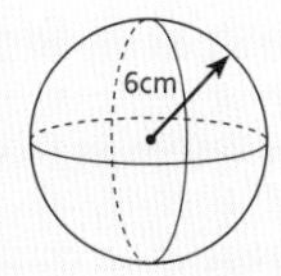


**a)** $V = \frac{4}{3} \times \pi \times 6^3 = 288\pi \text{cm}^3$

**b)** $SA = 4 \times \pi \times 6^2 = 144\pi \text{cm}^2$

## Composite Shapes

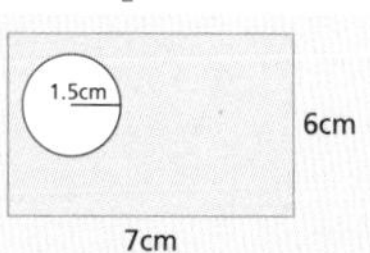


Calculate the area of the shaded region.

$A = (6 \times 7) - (\pi \times 1.5^2)$

$= 34.9\text{cm}^2$ (to 1 d.p.)

Find the area of the rectangle and subtract the area of the circle.

### Key Point

To find the volume of a composite shape, you must break the shape down.

Work out the volume of the shape.
Give your answers to 2 decimal places.

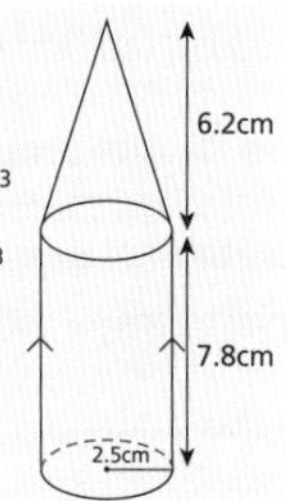


Volume of the Cylinder = $2.5^2 \times \pi \times 7.8 = 153.15\text{cm}^3$

Volume of the Cone = $\frac{1}{3} \times \pi \times 2.5^2 \times 6.2 = 40.58\text{cm}^3$

Total Volume = $153.15 + 40.58 = 193.73\text{cm}^3$

### Quick Test

1. Work out the volume of a sphere with diameter 10cm.
2. Calculate the surface area of a cone with radius 3cm and perpendicular height 6cm.
3. Work out the volume of a square-based pyramid with side length 5cm and perpendicular height 8cm.
4. Calculate the surface area of a hemisphere with radius 6cm.

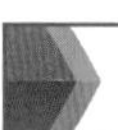

### Key Words

**pyramid**
**vertex / vertices**
**cone**
**frustum**
**similar**
**sphere**
**hemisphere**

# Review Questions

## Angles and Shapes 1 & 2

1. The three interior angles of a triangle are $y°$, $2y°$ and $3y°$.

   Work out the size of the largest angle. [2]

2. A quadrilateral has angles of 80°, 123°, 165° and 40°.

   Why is it impossible to draw this quadrilateral? [1]

3. Name all the quadrilaterals that can be drawn with four lines of length 5cm, 8cm, 5cm and 8cm. [2]

4. An aircraft flies from airport A on a bearing of 054° to airport B.

   Work out the bearing that the aircraft must fly on in order to return to airport A. [1]

5. What is the size of one interior angle in an equilateral triangle? [1]

6. Calculate:

   **a)** The sum of the interior angles in a regular octagon [2]

   **b)** The size of one interior angle in a regular octagon. [1]

7. A map has a scale of 1cm : 3km. A lake on the map is 6.5cm long.

   What is the actual length of the lake in kilometres? [1]

8. The angles in a triangle are $y + 5$, $3y - 16$ and $2y + 5$ degrees.

   **a)** Write down an equation for the sum of the angles in the triangle. Give your answer in its simplest form. [1]

   **b)** Solve your equation to find the value of $y$. [1]

   **c)** Work out the size of each angle in the triangle. [3]

9. Elaine calculates that the interior angle of a regular polygon is 158°. Pauline says that Elaine has made a mistake.

   Who is correct?
   Give a reason for your answer and show all your working. [3]

**Total Marks** ........ / 19

# Transformations, Constructions & Nets, Plans and Elevations

**1** **a)** Plot the following points: $A(2, 0)$ $B(5, 0)$ $C(5, 2)$ $D(3, 2)$ $E(3, 5)$ $F(2, 5)$
Join the points together and label the shape M. [1]

**b)** Rotate shape M by 180° about the origin to form shape N. [1]

**c)** Reflect shape N in the $x$-axis to form shape O. [1]

**d)** Describe fully the single transformation that maps shape O to shape M. [2]

**2** Rectangle R has a width of 3cm and a length of 5cm.
It is enlarged by scale factor 3 to give rectangle T.

**a)** What is the area of rectangle T? [2]

**b)** How many times bigger is the area of rectangle T than the area of rectangle R? [2]

**3** Describe how to construct an angle of 45°. [2]

**4** Describe the locus of points in the following:

**a)** A person sitting on the London Eye as it rotates around. [1]

**b)** The seat of a moving swing. [1]

**c)** The end of the minute hand on a clock moving for one hour. [1]

**d)** The end of a moving see-saw. [1]

**5** The diagram represents a solid made from ten identical cubes.

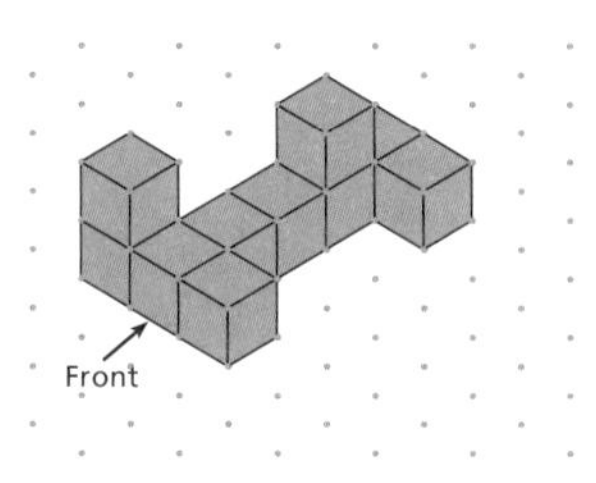


On a squared grid, draw the:

**a)** Front elevation **b)** Plan view. [2]

# Practice Questions

**6** Which of the following four nets produce cubes?
Circle your answers.

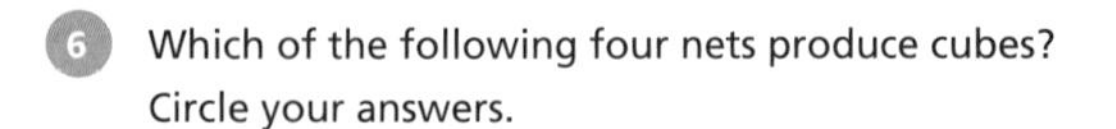

A B C D [1]

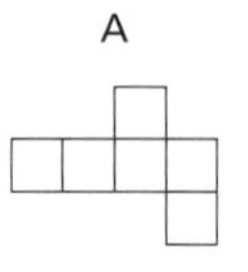
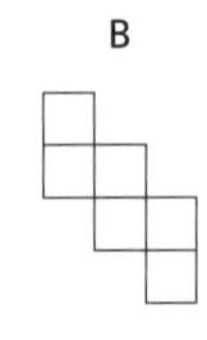
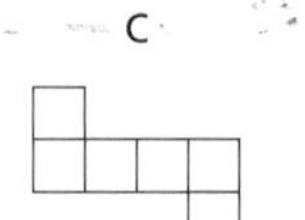
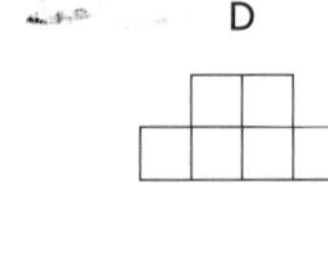

**7** What is the order of rotational symmetry of the following?

**a)** The letter H [1]

**b)** The letter Z. [1]

**8** **a)** On a set of axes, labelled from 0 to 10, plot the points $A(9, 6)$, $B(5, 10)$, $C(1, 6)$ and $D(1, 2)$.
Join the points together. [1]

**b)** On the same axes, plot the points $E(9, 3)$, $F(8, 4)$, $G(7, 3)$ and $H(7, 2)$.
Join the points together. [1]

**c)** Write down the transformation that maps shape $ABCD$ to shape $EFGH$. [3]

**9** The diagram shows a 3D shape.

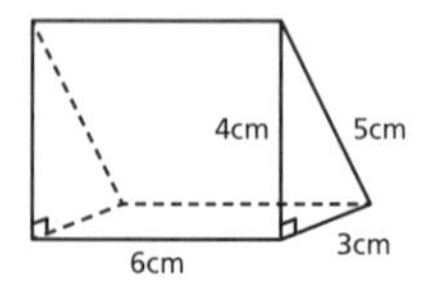


**a)** Draw an accurate net for the shape. [2]

**b)** How many faces does the shape have? [1]

**10** Here is the plan view, front elevation and side elevation of a 3D shape.

Front Side Plan

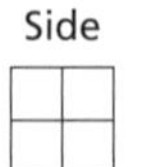

Sketch the 3D shape. [3]

**Total Marks** ........................ / 31

# Area and Volume 1, 2 & 3

**1** Work out:

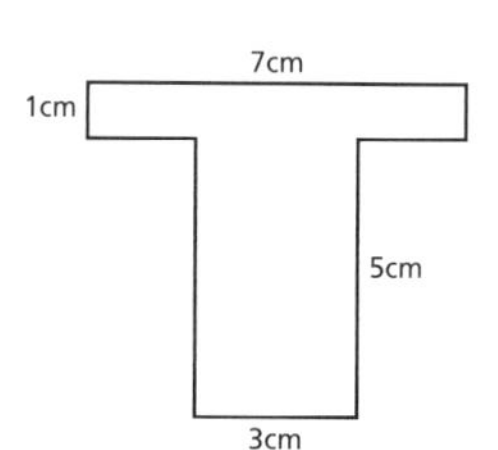


**a)** The perimeter of the shape [2]

**b)** The area of the shape. [2]

**2** Here is a cuboid. The volume of the cuboid is $300\text{cm}^3$.

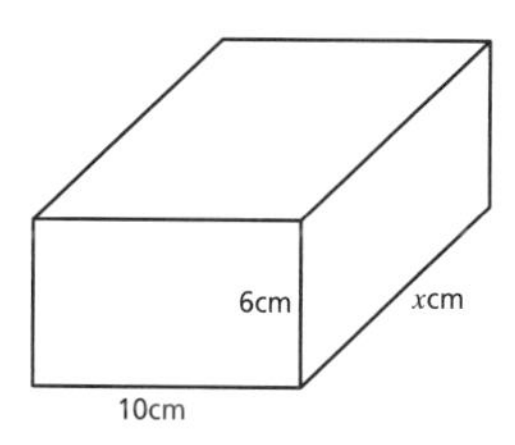


Work out the value of $x$. [2]

**3** Two identical circles sit inside a square of side length 6cm.

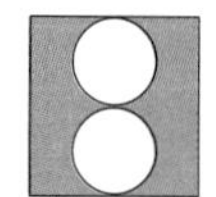

Work out the area of the shaded region. [4]

**4** A vase is made from two cylinders. The larger cylinder has a radius of 15cm.
The total volume of the vase is $6000\pi\text{cm}^3$.
The ratio of volumes of the smaller cylinder to the larger cylinder is 1 : 3.

**a)** Calculate the height of the larger cylinder. [3]

**b)** The height and radius of the smaller cylinder are equal.

Work out the radius of the smaller cylinder. [3]

**5** A cat's toy is made out of plastic. The top of the toy is a solid cone with radius 3cm and height 7cm. The bottom of the toy is a solid hemisphere. The base of the hemisphere and the base of the cone are the same size.

Calculate the volume of plastic needed to make the toy. Give your answer in terms of $\pi$. [3]

**Total Marks** ........................ / 19

# Congruence and Geometrical Problems

**You must be able to:**

- Identify congruent and similar shapes
- State the criteria that congruent triangles satisfy
- Solve problems involving similar figures
- Understand geometrical problems.

## Congruent Triangles

- If two shapes are the same size and shape, they are **congruent**.
- Two triangles are congruent if they satisfy one of the following four criteria:
  - SSS – three sides are the same
  - SAS – two sides and the **included angle** (the angle between the two sides) are the same
  - ASA – two angles and one corresponding side are the same
  - RHS – there is a right angle, and the hypotenuse and one other corresponding side are the same.
- Sometimes angles or lengths of sides have to be calculated before congruency can be proved.

State whether these two triangles are congruent and give a reason for your answer.

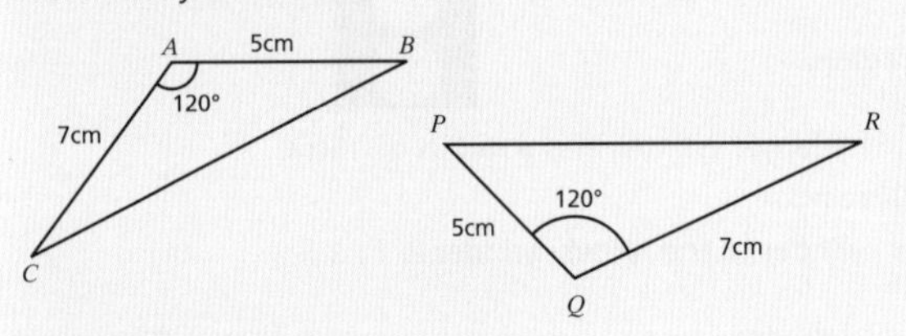


Angle $CAB$ = Angle $PQR$ (given)
$AC = QR$ (given)
$AB = PQ$ (given)
Triangles $ABC$ and $PQR$ are congruent because they satisfy the criteria SAS.

**Key Point**

Congruent shapes can be reflected, rotated or translated and remain congruent.

## Similar Triangles

- **Similar** figures are identical in shape but can differ in size.
- In similar triangles:
  - corresponding angles are identical
  - lengths of corresponding sides are in the same ratio $y : z$
  - the area ratio = $y^2 : z^2$
  - the volume ratio = $y^3 : z^3$.

Triangles $AED$ and $ABC$ are similar.

Calculate **a)** $AC$ and **b)** $DC$.

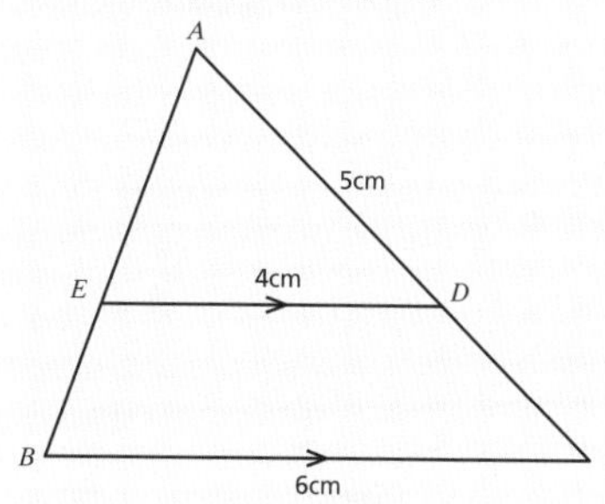


**a)** $\frac{4}{6} = \frac{5}{AC}$ ← The corresponding sides of both triangles are in the same ratio.

$4 \times AC = 6 \times 5$ ← Cross multiply.

$AC = \frac{6 \times 5}{4}$

$= 7.5\text{cm}$

**b)** $DC = AC - AD$

$= 7.5 - 5$

$= 2.5\text{cm}$

## Geometrical Problems

- Congruency and similarity are used in many geometric proofs.

Prove that the base angles of an isosceles triangle are equal.

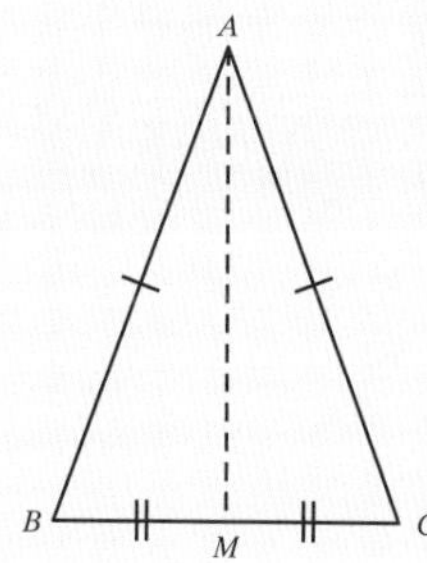


Given $\Delta\, ABC$ with $AB = AC$
Let $M$ be the midpoint of $BC$
Join $AM$
$AB = AC$ (given)
$BM = MC$ (from construction)
$AM = AM$ (common side)
$\Delta\, ABM$ and $\Delta\, ACM$ are congruent (SSS)
So, angle $ABC$ = angle $ACB$

### Key Point

When writing a proof, always give a reason for each statement.

### Quick Test

1.

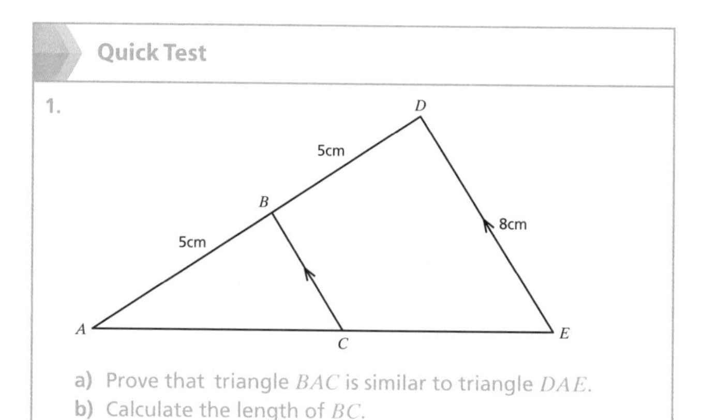


a) Prove that triangle $BAC$ is similar to triangle $DAE$.
b) Calculate the length of $BC$.

### Key Words

congruent
included angle
similar

# Right-Angled Triangles 1

**You must be able to:**

- Recall and use the formula for Pythagoras' Theorem
- Calculate the length of an unknown side in a right-angled triangle
- Apply Pythagoras' Theorem to real-life problems
- Use Pythagoras' Theorem in isosceles triangles.

## Pythagoras' Theorem

- The longest side ($c$) of a right-angled triangle is called the **hypotenuse.**
- **Pythagoras' Theorem** states that $a^2 + b^2 = c^2$.

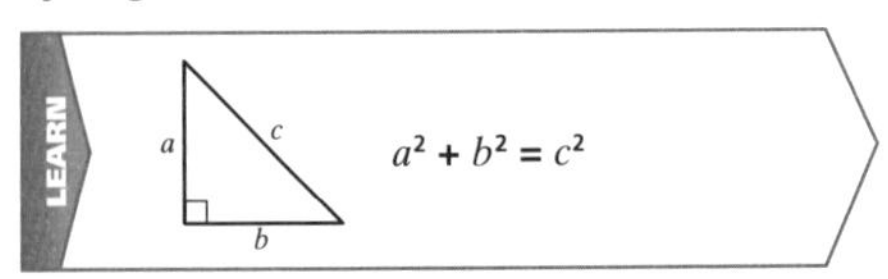

$a^2$ means $a \times a$ **not** $2 \times a$

## Calculating Unknown Sides

Calculate the length of side $c$.

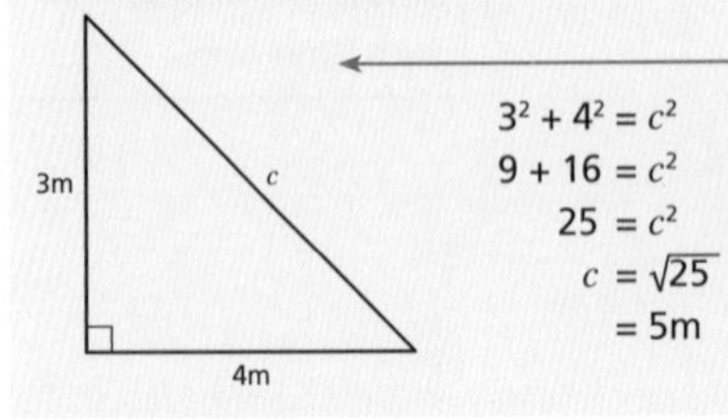


$$3^2 + 4^2 = c^2$$
$$9 + 16 = c^2$$
$$25 = c^2$$
$$c = \sqrt{25}$$
$$= 5\text{m}$$

$c$ is the hypotenuse (the longest side).

Calculate the length of side $b$.

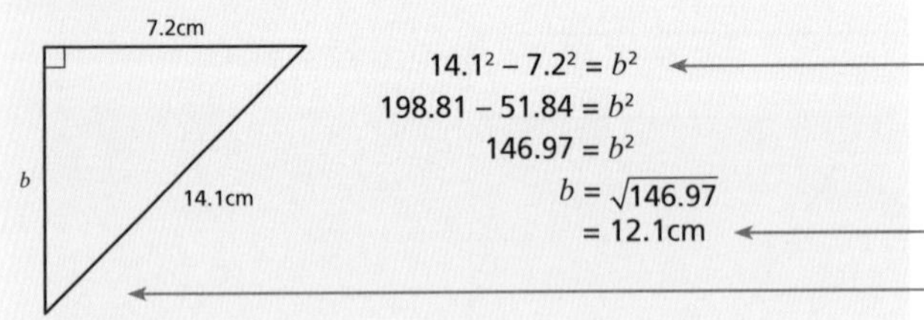


$$14.1^2 - 7.2^2 = b^2$$
$$198.81 - 51.84 = b^2$$
$$146.97 = b^2$$
$$b = \sqrt{146.97}$$
$$= 12.1\text{cm}$$

Rearrange the formula: $a^2 + b^2 = c^2 \rightarrow c^2 - a^2 = b^2$

Give the answer to 3 significant figures unless you are told otherwise.

$b$ is one of the shorter sides, so the answer must be less than 14.1cm.

- The following length combinations are **Pythagorean triples.** They regularly appear in right-angled triangles:
  - (3, 4, 5)
  - (6, 8, 10)
  - (5, 12, 13)
  - (7, 24, 25).

**Key Point**

Memorise the Pythagorean triples to help identify unknown sides quickly.

## Real-Life Problems

A boat sails 15km due north and then 10km due east.

How far is the boat from its starting point?
Give your answer to 3 decimal places.

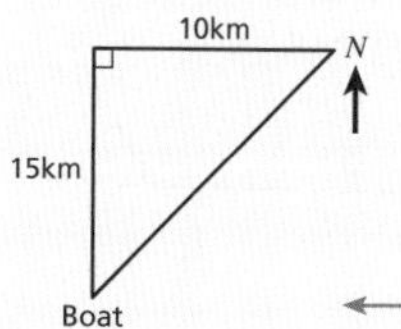


$$15^2 + 10^2 = c^2$$
$$225 + 100 = c^2$$
$$325 = c^2$$
$$c = \sqrt{325} = 18.028\text{km}$$

**Key Point**

A written problem might require a quick sketch to help you visualise the information and solve it.

A sketch makes it clear that you are looking for the hypotenuse ($c$) of a right-angled triangle.

A four-metre ladder leans against a tree.
It reaches three metres up the side of the tree.

Calculate how far the base of the ladder is from the bottom of the tree.

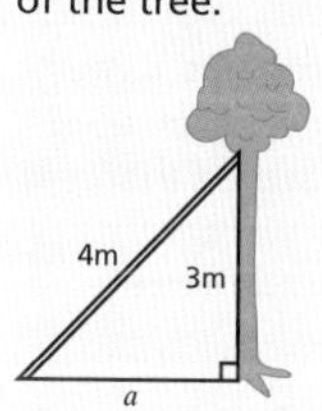


$$4^2 - 3^2 = a^2$$
$$16 - 9 = a^2$$
$$7 = a^2$$
$$a = \sqrt{7}$$
$$= 2.65\text{m}$$

You are looking for one of the shorter sides of a right-angled triangle.

## Isosceles Triangles

Calculate the height of the isosceles triangle $ABD$.

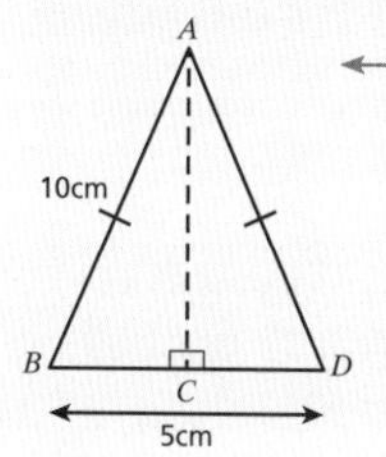


$$10^2 - 2.5^2 = AC^2$$
$$100 - 6.25 = AC^2$$
$$93.75 = AC^2$$
$$AC = \sqrt{93.75}$$
$$= 9.68\text{cm}$$

The height of the triangle is $AC$ (one of the short sides of a right-angled triangle).

$C$ is the midpoint of $BD$, so $BC = 5 \div 2 = 2.5$cm.

**Quick Test**

1. A rectangle measures 12cm by 5cm.
   Work out the length of its diagonal.
2. A triangle has sides 8cm, 15cm and 17cm.
   Show that it is a right-angled triangle.
3. A caterpillar was eating a cabbage. The wind blew the caterpillar 24cm due north. The wind suddenly changed direction and then blew the caterpillar 56cm due west. How far was the caterpillar from the cabbage?

**Key Words**

hypotenuse
Pythagoras' Theorem
Pythagorean triple

# Right-Angled Triangles 2

**You must be able to:**

- Recall and use the trigonometric ratios
- Calculate unknown lengths and angles using trigonometry
- Recall the exact trigonometric values for certain angles without using a calculator.

## The Trigonometric Ratios

- Unknown sides or angles in right-angled triangles can be calculated using the **trigonometric ratios: sine, cosine** and **tangent**.
- The symbol $\theta$ (**theta**) is used to represent an unknown angle.

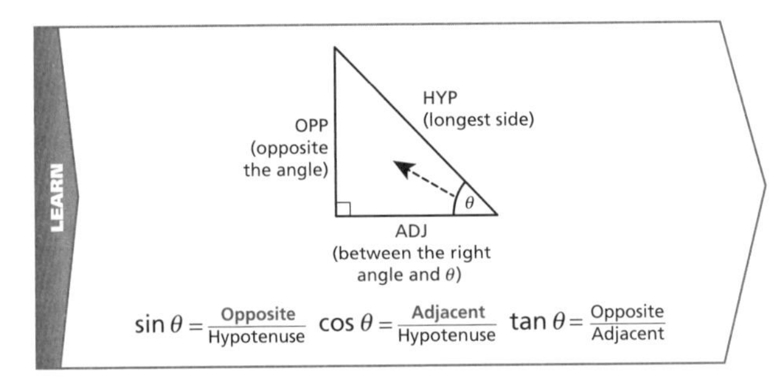


> **Key Point**
>
> Sine, cosine and tangent ratios can only be used in right-angled triangles.

- The above formulae can be remembered using:
  **S**ome **O**ld **H**orses **C**arry **A** **H**eavy **T**on **O**f **A**pples
  **SOH** **CAH** **TOA**

For example:
$\sin\theta\,(\textbf{S}\text{ome}) = \dfrac{\text{Opposite }(\textbf{O}\text{ld})}{\text{Hypotenuse }(\textbf{H}\text{orses})}$

## Calculating Unknown Sides

Work out the length of $x$. Give your answer to 1 decimal place.

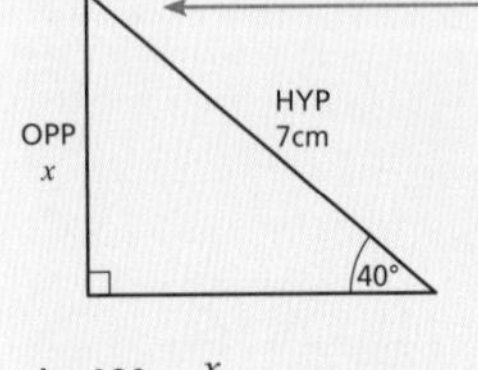


You have the hypotenuse (H) and you are looking for the side opposite (O) the given angle, so use sine (SOH).

$\sin 40° = \dfrac{x}{7}$

$x = 7 \times \sin 40°$

$x = 7 \times 0.6428$

$x = 4.499$

$x = 4.5\text{cm}$ (to 1 decimal place)

Rearrange, using inverse operations, to work out the value of $x$.

The diagram shows a lookout tower.
Ann is standing 30m from the base of the tower.

If the angle of elevation to the top of the tower is 50°, calculate the height of the tower. Give your answer to 1 decimal place.

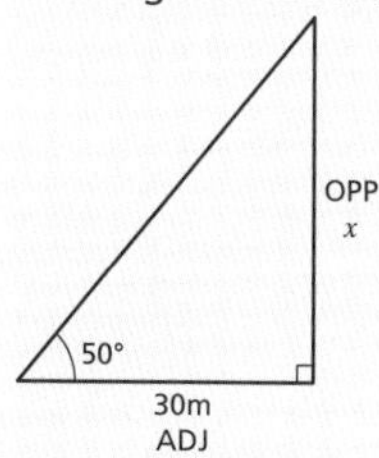


$$\tan 50° = \frac{x}{30}$$
$$x = 30 \times \tan 50°$$
$$x = 30 \times 1.1918$$
$$= 35.8\text{m (to 1 decimal place)}$$

You have the adjacent (A) side and you are looking for the side opposite (O) the given angle, so use tangent (TOA).

**Key Point**

Angles of elevation are angles above the horizontal, e.g. the angle from the ground to the top of a tower.

## Calculating Unknown Angles

Work out the size of angle $\theta$.
Give your answer to the nearest degree.

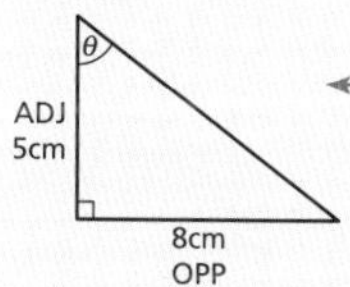


$$\tan \theta = \frac{8}{5} = 1.6$$
$$\tan^{-1} 1.6 = 57.99°$$
$$\theta = 58°$$

You have the adjacent (A) side and the opposite side (O), so use tangent (TOA).

On your calculator press SHIFT tan 1 · 6

$EFG$ is a right-angled triangle. Angle $EFG = 90°$.
$FG = 5\text{cm}$ and $EG = 8\text{cm}$.

Calculate angle $EGF$. Give your answer to 1 decimal place.

$$\cos \theta = \frac{5}{8} = 0.625$$
$$\cos^{-1} 0.625 = 51.3°$$

$EG$ is the hypotenuse (H) as it is opposite the right angle and $FG$ is the side adjacent (A) to the unknown angle, so use cosine (CAH).

## Trigonometric Values to Learn

| | sin | cos | tan |
|---|---|---|---|
| 0° | 0 | 1 | 0 |
| 30° | $\frac{1}{2}$ | $\frac{\sqrt{3}}{2}$ | $\frac{1}{\sqrt{3}}$ |
| 45° | $\frac{1}{\sqrt{2}}$ | $\frac{1}{\sqrt{2}}$ | 1 |
| 60° | $\frac{\sqrt{3}}{2}$ | $\frac{1}{2}$ | $\sqrt{3}$ |
| 90° | 1 | 0 | infinity |

**Key Point**

Make sure your calculator is in degree mode when using the trigonometric ratios.

**Quick Test**

1. An 8m ladder leans against a vertical wall. The base of the ladder is 3.5m from the wall. Calculate the angle between the top of the ladder and the wall. Give your answer to one decimal place.
2. A ship sails 14km on a bearing of 035°. How far north has the ship travelled? Give your answer to the nearest kilometre.

**Key Words**

**trigonometric ratios**
**sine**
**cosine**
**tangent**
**theta**
**opposite**
**adjacent**

# Review Questions

## Transformations, Constructions & Nets, Plans and Elevations

1. Three points $X(5,1)$, $Y(3, 5)$, and $Z(1, 2)$ are reflected in the $y$-axis.

   a) Give the new coordinates of the three points. [3]

   b) The original points $X$, $Y$, and $Z$ are rotated 90° about (0, 0) in a clockwise direction.

   Give the coordinates of the three points in their new positions. [3]

2. A rectangle (C) measures 3cm by 5cm. Each length of rectangle (C) is enlarged by scale factor 3 to form a new rectangle (D).

   What is the ratio of the area of rectangle C to rectangle D? [3]

3. A cuboid (C) measures 3cm by 4cm by 5cm. Each length of cuboid C is enlarged by scale factor 3 to form a new cuboid (D).

   What is the ratio of the volume of cuboid C to cuboid D? [3]

4. On a 6 × 6 grid, plot the points $A(3, 2)$, $B(1, 3)$, $C(0, 6)$ and $D(2, 5)$.

   Reflect each point in the line that joins (3, 0) to (3, 6) and write down the coordinates of points $A'$, $B'$, $C'$ and $D'$ in the image produced. [4]

5. Describe the locus of points for the following:

   a) The path of a rocket for the first three seconds after take-off. [1]

   b) A point just below the handle on an opening door. [1]

   c) The central point of a bicycle wheel as the bicycle travels along a level road. [1]

   d) The end of a pendulum on a grandfather clock. [1]

6. Describe the plan view of a cube measuring 4cm by 4cm by 4cm. [1]

7. a) Construct a triangle, $DEF$, where $DE$ = 8cm, $EF$ = 7cm and $DF$ = 3cm. [2]

   b) By accurate measurement, find the size of angle $FDE$. [1]

   c) Construct the bisector of angle $FED$. [2]

8. The photograph shows a World War II Lancaster Bomber.

Sketch:

a) the side elevation of the Lancaster Bomber [2]

b) the front elevation of the Lancaster Bomber [2]

c) the plan view of the Lancaster Bomber. [2]

9. Below is a 3D shape made up of eight cubes.

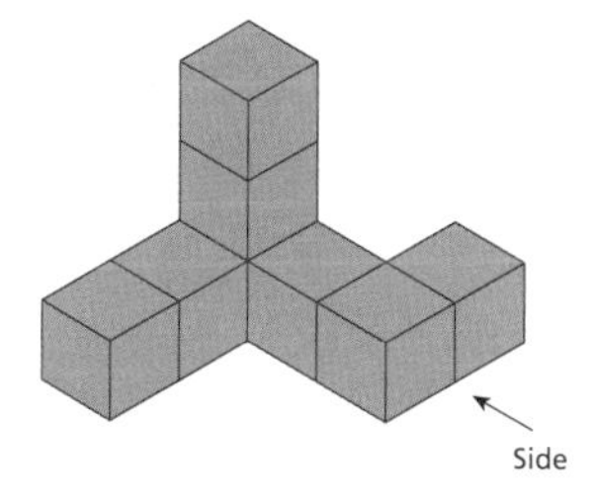


a) Draw the plan view. [2]

b) Draw the side elevation. [2]

10. a) Here is the top half of a word. The dashed line is the line of symmetry.

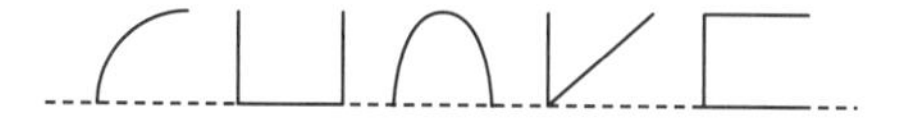

Write down the word. [1]

b) Give an example of another three or four-letter word, through which a horizontal line of symmetry can be drawn. [1]

Total Marks ........ / 38

# Review Questions

## Area and Volume 1, 2 & 3

**1** **a)** Work out the volume of the triangular prism. [2]

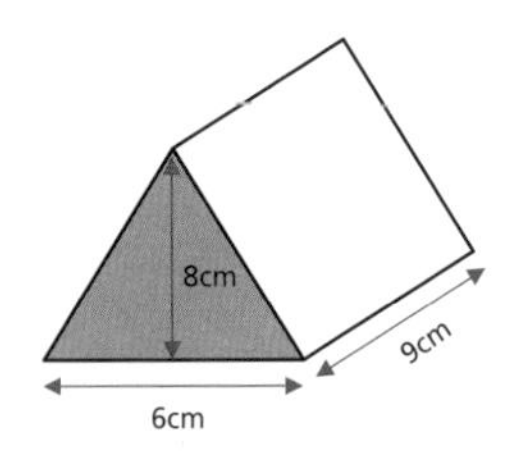


**b)** A cube has the same volume as the triangular prism.

Work out the total length of all the edges of the cube. [3]

**2** The numerical values of the area and circumference of a circle are equal.

Work out the radius of this circle. [2]

**3** The volume of the trapezoid is $900\text{cm}^3$.
All measurements are in centimetres.

Work out the value of $x$. [4]

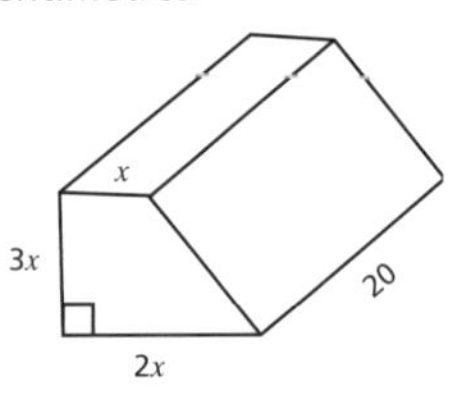


**4** The surface area of a sphere is $75\text{cm}^2$. Work out the length of the radius. [3]

**5** Here is a triangle.
The area of the triangle is $7.5\text{cm}^2$.

Work out the value of $x$. [3]

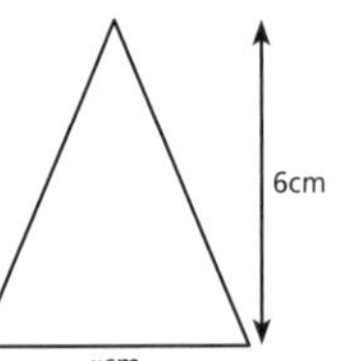


**6** John is planning to paint the front of his house. He needs to estimate how much paint he should buy. He does this by calculating the area of the front of the house, including all windows and doors.

The diagram shows John's house.

If each tin of paint will cover $11\text{m}^2$, work out an estimate of the number of tins that John needs to buy. [4]

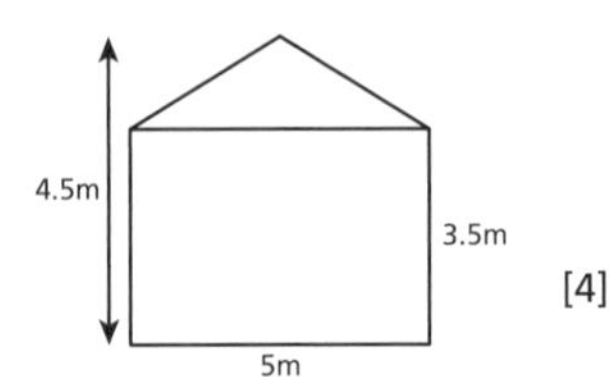


**Total Marks** ______ / 21

# Area and Volume 1, 2 & 3

**1** A rectangle has a length of 6cm and area of $24cm^2$.

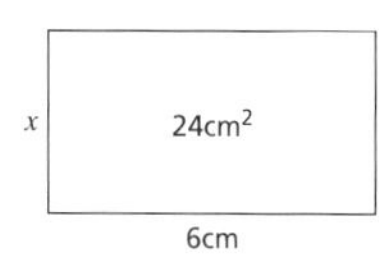


**a)** Work out the width of the rectangle ($x$cm).

Answer .................................... [2]

**b)** Calculate the perimeter of the rectangle.

Answer .................................... [2]

**2** A square has a side length of 6cm.

An equilateral triangle has the same perimeter as the square.

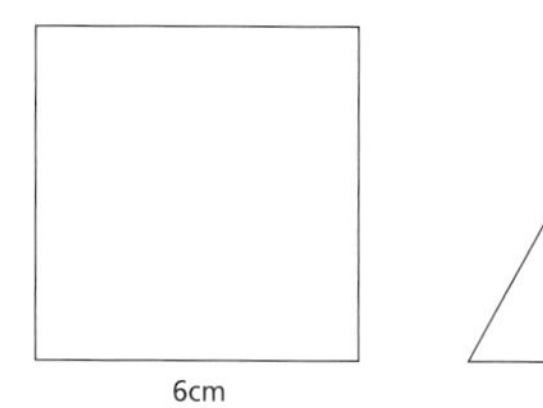


Work out the length of one side of the triangle.

Answer .................................... [3]

**Total Marks** ........................ / 7

# Practice Questions

## Congruence and Geometrical Problems

**1** Prove that triangle $ABC$ and triangle $BCD$ are similar. [3]

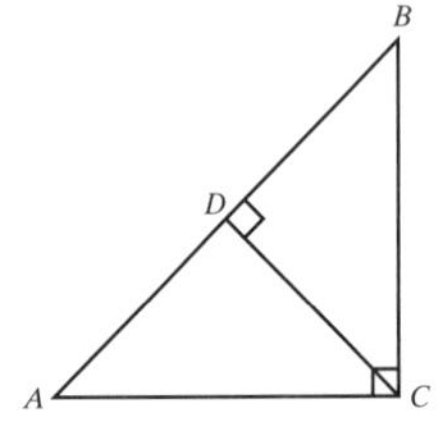


**2** Lisa has a 10cm by 8cm photograph of her pet dog. She wants a smaller copy to fit into her handbag and a larger copy for her office.

**a)** What will the length of the smaller copy be, if the width is 4cm? [1]

**b)** What will the width of the larger copy be, if the length is 25cm? [2]

**3**

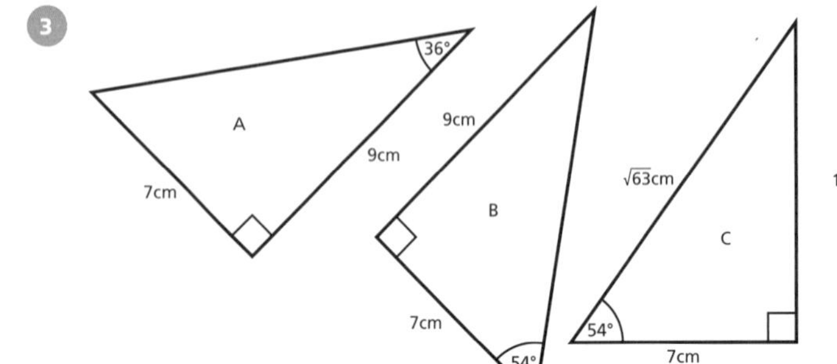


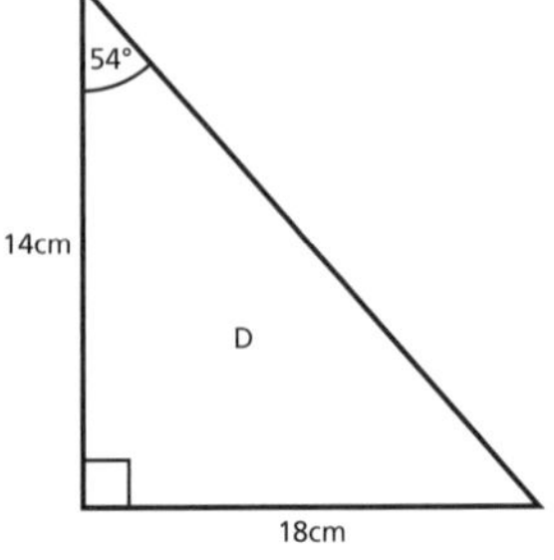


**a)** Which two triangles are congruent?
Give a reason for your answer.

Answer .................................... [2]

**b)** Which triangles are similar to triangle A?

Answer .................................... [1]

**Total Marks** .................... / 9

# Right-Angled Triangles 1 & 2

1. A man walks 6.7km due north. He then turns due west and walks 7.6km.

   How far is he now from his starting point? [3]

2. A 4m ladder leans against a vertical wire fence. The foot of the ladder is 2m from the base of the fence. Fang, the lion, can jump 3m vertically.

   Will Fang be able to jump over the fence?
   You must give reasons for your answer. [4]

3. $ABC$ is an isosceles triangle. $AB = BC =$ 13cm. $D$ is the midpoint of $AC$ and $AC =$ 10cm.

   Calculate the length of $BD$. [3]

4. A bumblebee leaves its nest and flies 10 metres due south and then 6 metres due west.

   What is the shortest distance the bumblebee has to fly to return to its nest?
   Give your answer to 3 significant figures. [3]

5. A triangle has side lengths of 1.5cm, 2.5cm and 2cm. Is it a right-angled triangle?
   Give a reason for your answer. [3]

6. How long is the diagonal of a square of side length 3cm? [2]

7. $A$ is the point (4, 0) and $B$ is the point (7, 5).

   Calculate the angle between line $AB$ and the $x$-axis to the nearest degree. [2]

8. Molly cycles 5km in a north-easterly direction from Apton ($A$) to Bray ($B$). She then cycles 8km in a north-westerly direction from Bray to Chart ($C$).

   **a)** How far is Chart from Apton? Give your answer to 2 significant figures. [2]

   **b)** Calculate angle $CAB$. Give your answer to the nearest degree. [2]

9. 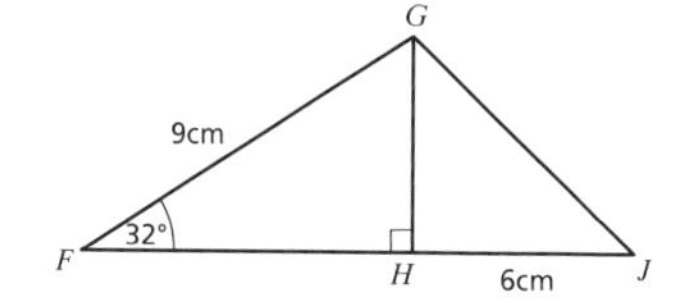


   Calculate:

   **a)** The length of $GH$.
   Give your answer to 2 decimal places. [2]

   **b)** The size of angle $HGJ$.
   Give your answer to 1 decimal place. [2]

**Total Marks** ........................ / 28

# Circles

You must be able to:

- Identify the parts of a circle and understand their basic properties
- Recognise cyclic quadrilaterals
- Calculate angles in a circle.

## Parts of a Circle

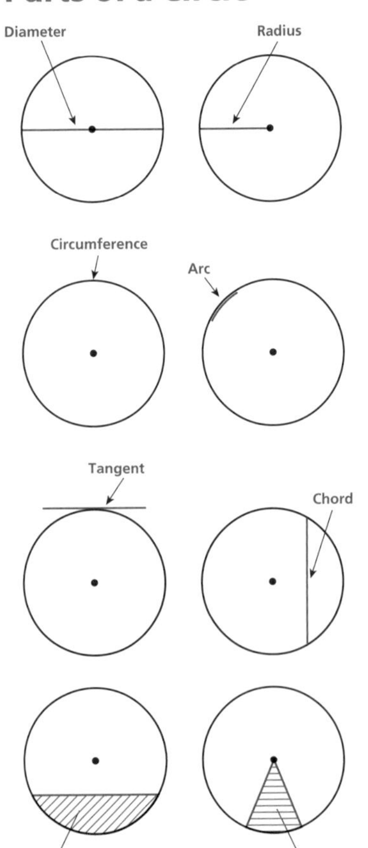

**Key Point**

The plural of radius is radii.

The radius of a circle is half the diameter.

**Key Point**

Half a circle is called a semicircle.

**Key Point**

Sector area is a fraction of the area of a circle.

Arc length is a fraction of the circumference of a circle.

- The circumference is found with the formula:

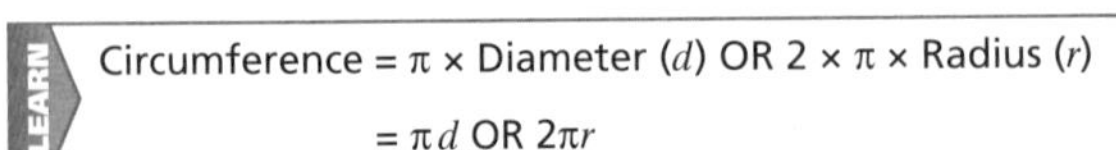


Circumference = $\pi \times$ Diameter ($d$) OR $2 \times \pi \times$ Radius ($r$)

$= \pi d$ OR $2\pi r$

## Cyclic Quadrilaterals

- All four vertices (corners) of a **cyclic quadrilateral** touch the circumference of the circle.
- The **opposite angles** of a cyclic quadrilateral **add up to 180°.**

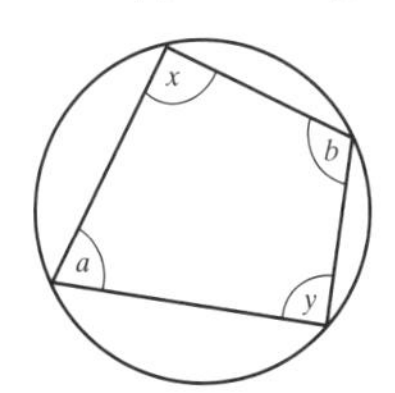


$a + b = 180°$
$x + y = 180°$

## Angles in a Circle

$O$ is the centre of the circle.
$AO$ and $BO$ are radii.

If angle $AOB = 80°$, calculate angle $ABO$.

$AO = BO$, so $AOB$ is an isosceles triangle.

Angle $ABO = \dfrac{(180° - 80°)}{2} = 50°$

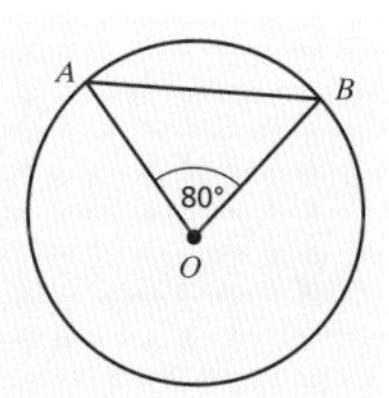


$O$ is the centre of the circle.
$AO$ and $BO$ are radii.

If angle $OAB = 54°$, calculate angle $AOB$.

$AO = BO$, so $AOB$ is an isosceles triangle.
Angle $OAB = OBA$
Angle $AOB = 180° - (54° + 54°) = 72°$

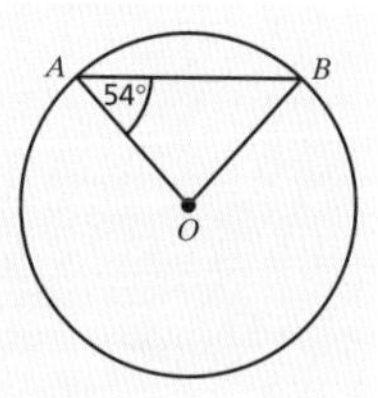


$O$ is the centre of the circle.

If angle $EOF = 93°$, and angle $EOG = 132°$, calculate angle $FOG$.

Angle $FOG = 360° - (132° + 93°) = 135°$

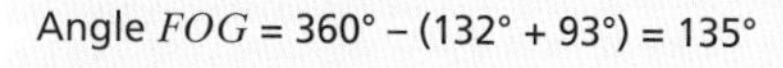

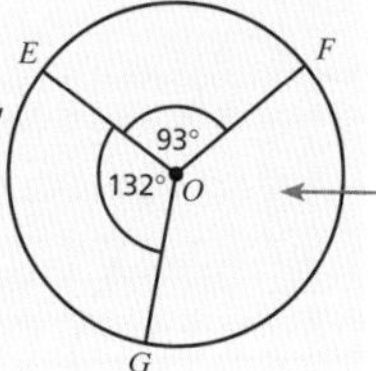


Angles at a point have a sum of 360°.

### Key Words

diameter
radius / radii
circumference
arc
tangent
chord
segment
sector
cyclic quadrilateral

### Quick Test

1. $O$ is the centre of a circle. $OX$ and $OY$ are radii. If angle $OYX = 40°$, calculate angle $XOY$.
2. Draw a circle with a radius of 4cm.

# Vectors

**You must be able to:**

- Add and subtract vectors
- Multiply a vector by a scalar
- Work out the magnitude of a vector
- Carry out translations according to column vectors.

## Properties of Vectors

**Key Point**

The direction of a vector is shown by an arrow.

- A **vector** is a quantity that has both **magnitude** (size) and direction.
- Vectors are equal only when they have equal magnitudes and are in the same direction.

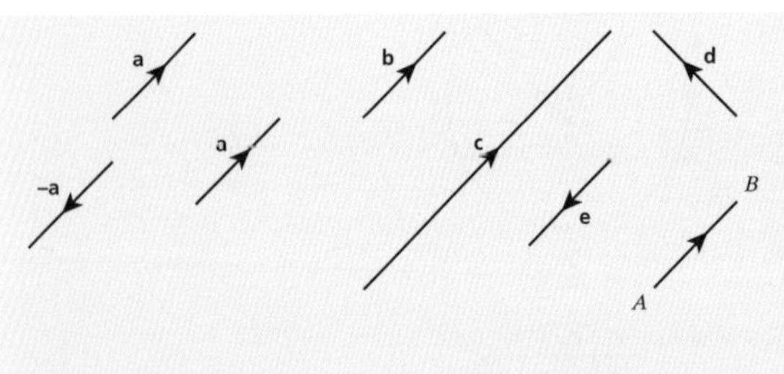


$\mathbf{b} = \mathbf{a}$ (same direction, same length)

$\mathbf{d} \neq \mathbf{a}$ (different direction, same length)

$\mathbf{e} = -\mathbf{a}$ (opposite direction, same length)

$\mathbf{c} = 3\mathbf{a}$ (same direction, $3 \times$ length of **a**)

$\mathbf{a} = \overrightarrow{AB} = \underline{a} = \begin{pmatrix} 2 \\ 2 \end{pmatrix}$ ← These are all ways of writing the same vector.

$-\mathbf{a} = \begin{pmatrix} -2 \\ -2 \end{pmatrix}$

**a** and **c** are parallel vectors.

**a** and **b** are equal vectors.

- Any number of vectors can be added together.

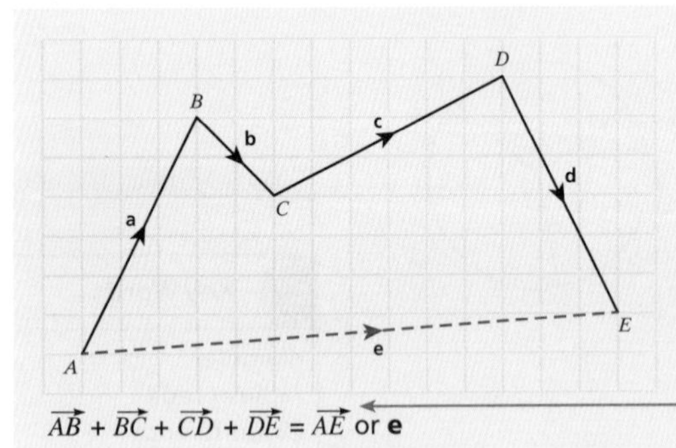


$$\mathbf{a} + \mathbf{b} + \mathbf{c} + \mathbf{d} = \begin{pmatrix} 3 \\ 6 \end{pmatrix} + \begin{pmatrix} 2 \\ -2 \end{pmatrix} + \begin{pmatrix} 6 \\ 3 \end{pmatrix} + \begin{pmatrix} 3 \\ -6 \end{pmatrix} = \begin{pmatrix} 14 \\ 1 \end{pmatrix}$$

$\overrightarrow{AB} + \overrightarrow{BC} + \overrightarrow{CD} + \overrightarrow{DE} = \overrightarrow{AE}$ or **e** ← This is the resultant vector.

- When a vector is multiplied by a **scalar** (a numerical value), the resultant vector will always be parallel to the original vector.
- When a vector is multiplied by a positive number (not 1), the direction of the vector does not change, only its magnitude.

**Key Point**

The sum of the lengths $AB + BC + CD + DE$ does **not** equal the length of $AE$.

- When a vector is multiplied by a negative number (not –1), the magnitude of the vector changes and the vector points in the opposite direction.
- The magnitude of a vector **a** is written |**a**|
- Magnitude of vector $\begin{pmatrix} x \\ y \end{pmatrix}$ is $\sqrt{x^2 + y^2}$

Work out the magnitude of vector **a**.

Vector **a** = $\begin{pmatrix} 5 \\ -8 \end{pmatrix}$

$a^2 = 5^2 + 8^2$

Use Pythagoras' Theorem.

$a^2 = 25 + 64$

$|a| = \sqrt{89}$

$|a| = 9.43$ (to 3 significant figures)

## Translations

- When a shape is translated, it does not change size or rotate. It moves left or right and up or down.
- The translation is represented by a column vector $\begin{pmatrix} x \\ y \end{pmatrix}$
- $x$ represents the distance moved **horizontally**: **positive** means to the **right**, **negative** means to the **left**.
- $y$ represents the distance moved **vertically**: **positive** means **up**, **negative** means **down**.

Translate the shaded shape by vector $\begin{pmatrix} -4 \\ -3 \end{pmatrix}$

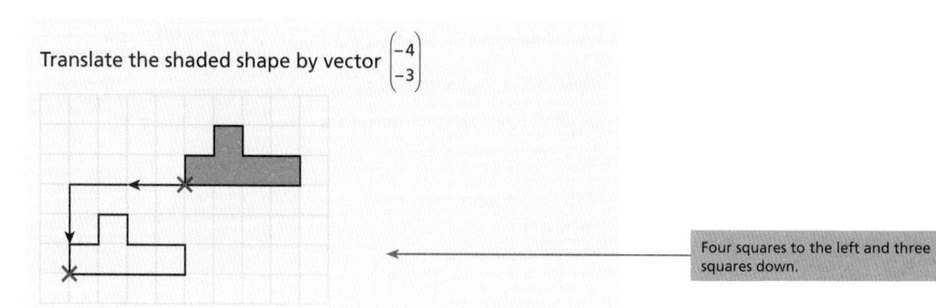

Four squares to the left and three squares down.

### Quick Test

1. Vector **a** has a magnitude of 3cm and a direction of 120°. Vector **b** has a magnitude of 4cm and a direction of 040°. Draw the vectors:
   a) **a** b) **b** c) –**a** d) 2**a** e) **a** + **b**

### Key Words

vector
magnitude
scalar

## Congruence and Geometrical Problems

1. A triangle has angles of 56°, 64° and 60°. The triangle is enlarged by scale factor 2.

   What are the angles of the enlarged triangle? [3]

2. A tree of height 5m casts a shadow that is 8.5m in length.

   Work out the height of a tree casting a shadow that is 34m in length. [2]

3. What is the difference between congruency and similarity? [2]

4. Fill in the missing criteria to complete the sentence:

   Two triangles are congruent if they satisfy one of four criteria: SSS, RHS, SAS and ................ . [1]

5. A tree that is 4m high casts a shadow 9.5m long.

   Work out the height a lamp post that casts a shadow 38m long.

   Answer ................................ [2]

6. A rectangle (A) has a length of 12cm and a width of 7cm. A similar rectangle (B) has a perimeter of 114cm.

   By what scale factor would rectangle A need to be enlarged to produce rectangle B?

   Answer ................................ [2]

Total Marks ................ / 12

## Right-Angled Triangles 1 & 2

1. A rectangle has a length of 10cm and a width of 5cm.

   Calculate the length of its diagonal. Give your answer to 3 significant figures. [2]

2. A square has a diagonal length of 12cm.

   Calculate the side length of the square. Give your answer to 2 decimal places. [3]

3. $PQR$ is a triangle. $P$ is at (1, 0), $Q$ is at (1, 5) and $R$ is at (6, 0).

   **a)** What type of triangle is $PQR$? [1]

   **b)** Calculate the length of $PR$. [1]

   **c)** Calculate the length of $PQ$. [1]

   **d)** Calculate the length of $QR$. Give your answer to 2 decimal places. [3]

   **e)** Work out the area of triangle $PQR$. [2]

4. Sean and Alexander are arguing about a triangle that has side lengths of 9cm, 40cm and 41cm. Sean says it is a right-angled triangle and Alexander says it is not.

   Who is correct? Write down a calculation to support your answer. [3]

5. Moira is standing 80m from the base of Blackpool Tower. The angle of elevation to the top of the tower is 63.15°.

   Calculate the height of the tower to the nearest metre. [3]

6. The coordinates of a triangle $XYZ$ are $X$(1, 4), $Y$(1, 1) and $Z$(6, 1).

   Calculate angle $XZY$. Give your answer to 3 decimal places. [2]

7. Chevaun says that sin 30° + sin 60° > cos 30° + cos 60°.

   Is she correct? Show working to support your answer. [3]

8. A helicopter leaves an air base in London and flies 175km in a north-easterly direction to an air base in Norwich.

   **a)** How far north of the London air base is the Norwich air base?
   Give your answer to the nearest kilometre. [3]

   **b)** How long did it take the helicopter to make the journey if it travelled 100km in 30 minutes? [2]

**Total Marks** ........ / 29

# Practice Questions

## Circles

**1** For each of the following questions, work out the lettered angles.
The centre of each circle is marked with an $O$ where appropriate.  [6]

**a)**
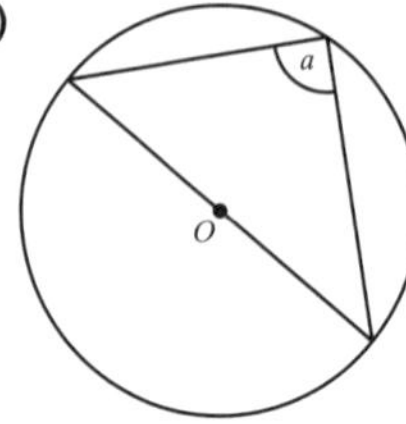


**b)**
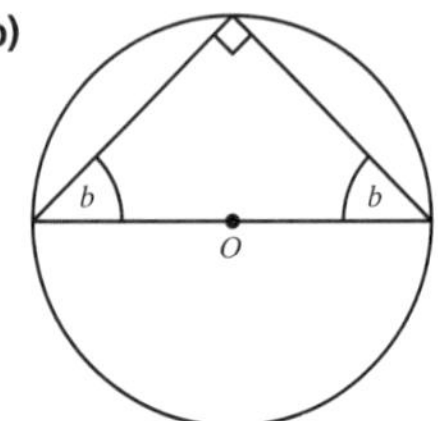


**c)**
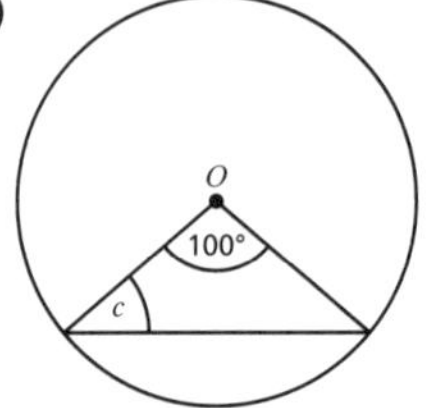


**d)**
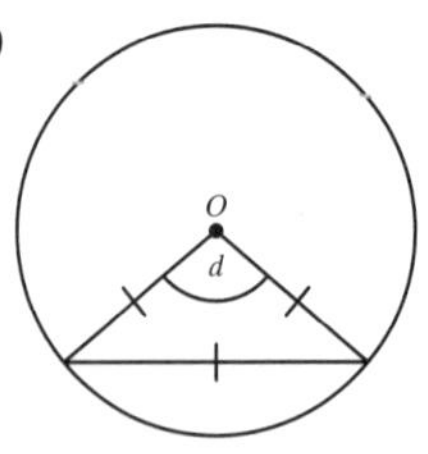


**e)**
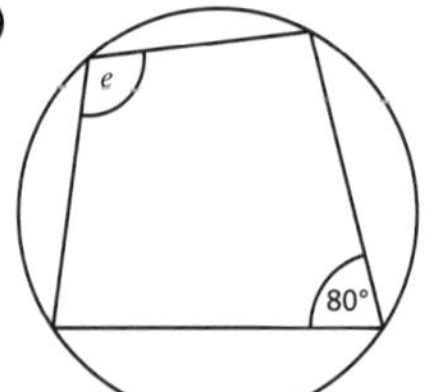


**f)**
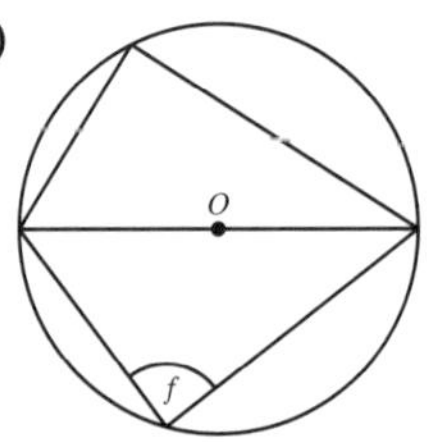


**Total Marks** ........ / 6

## Vectors

**1** On squared paper, draw a set of axes that go from –7 to +7 in each direction.

**a)** Plot the points $A(1, 2)$, $B(5, 2)$ and $C(5, 0)$ and join them together. Label the shape E.

What shape have you drawn? [2]

**b)** Translate shape E by the vector $\begin{pmatrix} 0 \\ -4 \end{pmatrix}$. Label the shape F.

What are the coordinates of shape F? [3]

**c)** Translate shape E by the vector $\begin{pmatrix} -7 \\ 3 \end{pmatrix}$. Label the shape G.

What are the coordinates of shape G? [3]

**Total Marks** ........ / 8

## Circles & Vectors

**1** Complete the sentences:

**a)** A curved line that is part of the circumference of a circle is an ____________. [1]

**b)** A straight line from the centre of the circle to the circumference is a ____________. [1]

**c)** The distance around the edge of the circle is the ____________. [1]

**d)** A line through the centre of the circle, with both ends touching the circumference, is the ____________. [1]

**e)** A line that touches the circumference at both ends, but does not pass through the centre, is a ____________. [1]

**f)** A line outside the circle, which touches the circumference of the circle at only one point, is a ____________. [1]

**g)** ____________ is the number 3.142 (to 3 decimal places). [1]

**h)** The formula used to calculate the area of a circle is ____________. [1]

**i)** The formula used to calculate the circumference of a circle is ____________. [1]

**2** Which of these vectors will be parallel to $4\mathbf{a} + 3\mathbf{b}$? Circle your answer.

$4\mathbf{a} + 6\mathbf{b}$ $8\mathbf{a} + 3\mathbf{b}$ $8\mathbf{a} + 6\mathbf{b}$ $4\mathbf{a} - 3\mathbf{b}$ [1]

**3** Translate shape A by the vector $\begin{pmatrix} -8 \\ 1 \end{pmatrix}$ [1]

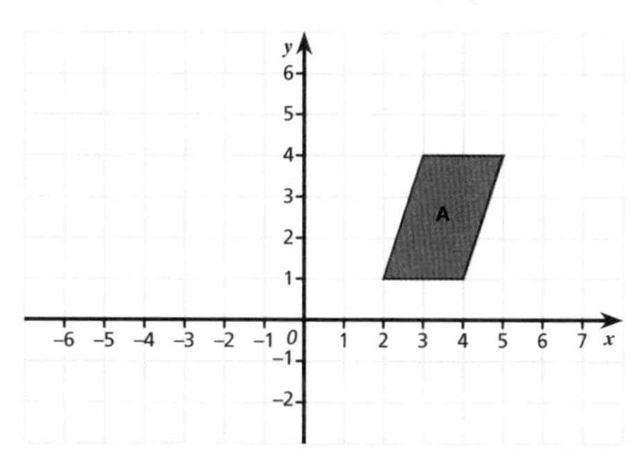


**Total Marks** ____________ / 11

# Answers

## Pages 4–7 Revise Reading

**Page 5 Quick Test**

1. **a)** Rectangle, Parallelogram, Kite
   **b)** Square, Rhombus
2. Angle $EJH = 16°$

> When $FE$ and $GH$ are made longer, they produce a triangle, $FJG$. Angles $EHF$ and $HFG$ are alternate angles on parallel lines.

**Page 7 Quick Test**

1. **a)** 3240°
   **b)** 162°
2. 12
3.

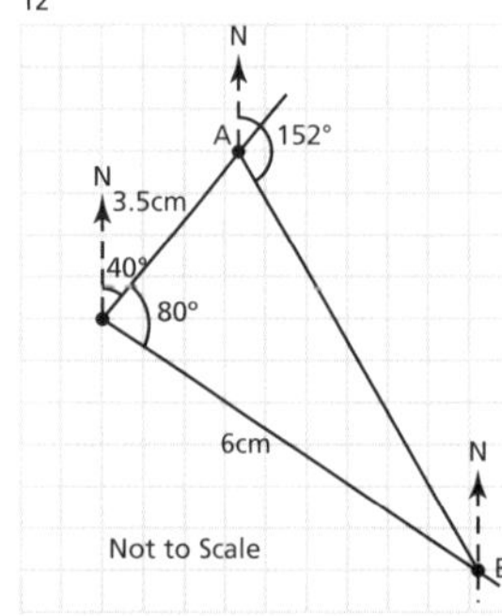


Bearing = 152° ± 2°

## Pages 8–9 Practice Questions

**Page 8**

1. (Marks will not be awarded if reason is incorrect.) $j = 72°$ (alternate angle) **[1]**; $k = 54°$ (sum of the interior angles of a triangle = 180°) **[1]**; $l = 54°$ (vertically opposite angles are equal) **[1]**; $m = 18°$ (90° – 72°) **[1]**
2. 360° – 46° – 107° – 119° = 88° **[1]**
3. Exterior angle = $\frac{360}{n} = \frac{360}{10} = 36°$ **[1]**
4. Bearing = 180° + 36° = 216° **[1]**
5. 36km ÷ 4 = 9, 9cm **[1]**
6. 315° **[1]**

**Page 9**

1. $9x = 360°$ **[1]**; $x = 40°$ **[1]**
2. Exterior angle (= 180° – 150°) = 30° **[1]**; number of sides = 360 ÷ 30 = 12 **[1]**
3. **a)** Correct scale drawing (see sketch) **[1]**; distance = 62km (+/– 2km) **[1]**

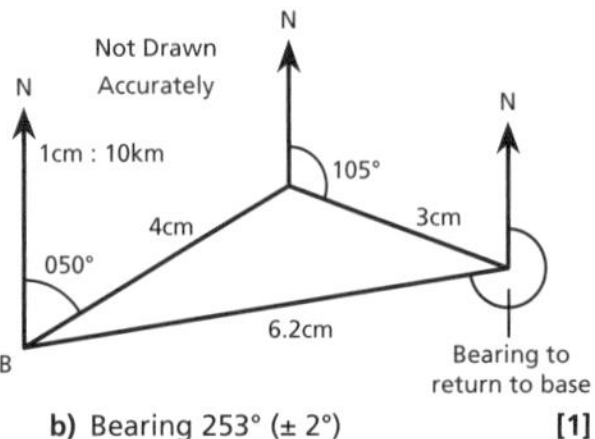


   **b)** Bearing 253° (± 2°) **[1]**

## Pages 10–21 Revise Reading

**Page 11 Quick Test**

1. **a)** A translation by the vector $\begin{pmatrix} 5 \\ -1 \end{pmatrix}$
   **b)** A reflection in the line $y = x$
   **c)** A rotation of 90° anticlockwise about (0, 0)

**Page 13 Quick Test**

1. Construct an angle of 60°and bisect it, i.e. draw a line and mark on it two points, $A$ and $B$.
   Open compasses to length $AB$.
   Put compass point on $A$ and draw an arc. Put compass point on $B$ and draw an arc.
   Draw a line to join $A$ to the new point, $C$.
   Adjust compasses so less than length $AB$.
   Put compass point on $A$ and draw arcs crossing $AB$ and $AC$ at points $D$ and $E$.
   Put compass point on $D$ and draw an arc. Put compass point on $E$ and draw an arc. Draw a line from $A$ to the new point, $F$.
2.

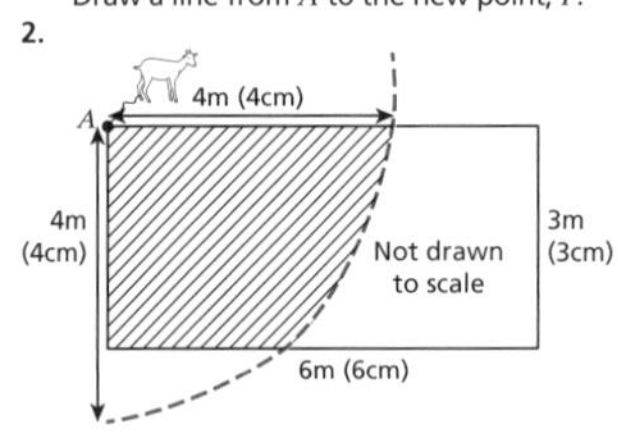


**Page 15 Quick Test**

1. 7cm, 7cm, 10cm

**Page 17 Quick Test**

1. Area = 12m², perimeter = 14m
2. 10cm²
3. Volume = 18cm³, surface area = 42cm²

**Page 19 Quick Test**

1. Volume = 301.59cm³ (to 2 d.p.) and surface area = 251.33cm² (to 2 d.p.) OR 301.44cm³ and 251.2cm² if using $\pi = 3.14$
2. 24cm²
3. Circumference = 21.99cm (to 2 d.p.) and area = 38.48cm² (to 2 d.p.) OR 21.98cm and 38.47cm² if using $\pi = 3.14$

**Page 21 Quick Test**

1. $\frac{500\pi}{3}$ or 523.6cm³
2. 91.5cm²
3. $\frac{200}{3}$ or 66.7cm³
4. $108\pi$ or 339.3cm²

## Page 22 Review Questions

**Page 22**

1. $y + 2y + 3y = 180°$, $6y = 180°$, $y = 30°$ **[1]**; Largest angle = 90° **[1]**
2. 80° + 123° + 165° + 40° = 408°
   It is not quadrilateral, because the interior angles are greater than 360° in total **[1]**
3. Kite, rectangle, parallelogram **[2]** (1 mark for two or three correct)
4. Bearing = 180° + 054° = 234° **[1]**
5. 60° **[1]**
6. **a)** Sum = 6 × 180° **[1]**; 1080° **[1]**
   **b)** $\frac{1080°}{8} = 135°$ **[1]**
7. 3 × 6.5 = 19.5km **[1]**
8. **a)** $y + 5 + 3y - 16 + 2y + 5 = 180°$, $6y - 6 = 180°$ **[1]**
   **b)** $6y = 186°$, $y = 31°$ **[1]**
   **c)** $y + 5 = 36°$ **[1]**; $3y - 16 = 77°$ **[1]**; $2y + 5 = 67°$ **[1]**
9. Pauline is correct **[1]**; exterior angle is 180° – 158° = 22° **[1]**; number of sides = 360° ÷ 22 = 16.36; it cannot be right because the number of sides must be a whole number **[1]**

## Pages 23–25 Practice Questions

**Page 23**

1.

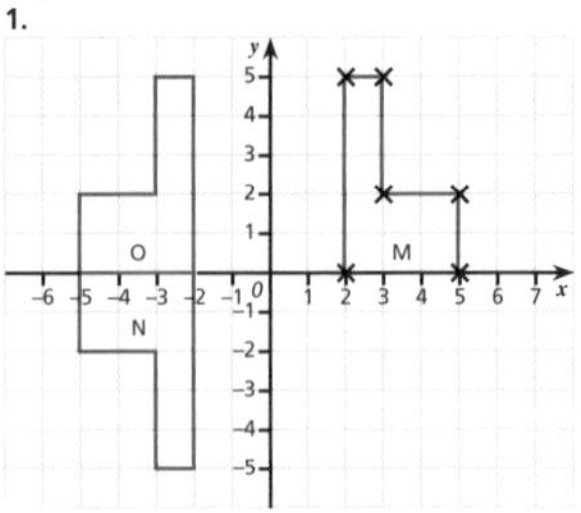


   **a)** Shape M plotted correctly **[1]**
   **b)** Shape N plotted correctly **[1]**
   **c)** Shape O plotted correctly **[1]**
   **d)** Reflection **[1]**; in the $y$-axis OR mirror line $x = 0$ **[1]**
2. **a)** Rectangle T is 9cm × 15cm **[1]**; Area = 135cm² **[1]**

b) Area R = 15cm², Area T = 135cm² **[1]**; T is 9 times bigger. **[1]**

3. Draw a line and construct the perpendicular bisector of the line. **[1]**; Bisect the right angle. **[1]**
4. a) A circle **[1]**
   b) An arc of a circle **[1]**
   c) A circle **[1]**
   d) An arc of a circle **[1]**
5. a) Front Elevation **[1]**
   b) Plan View

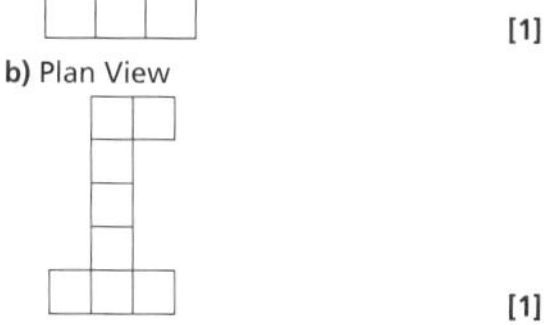

**[1]**

6. A, B and C **[1]**
7. a) 2 **[1]**
   b) 2 **[1]**
8.

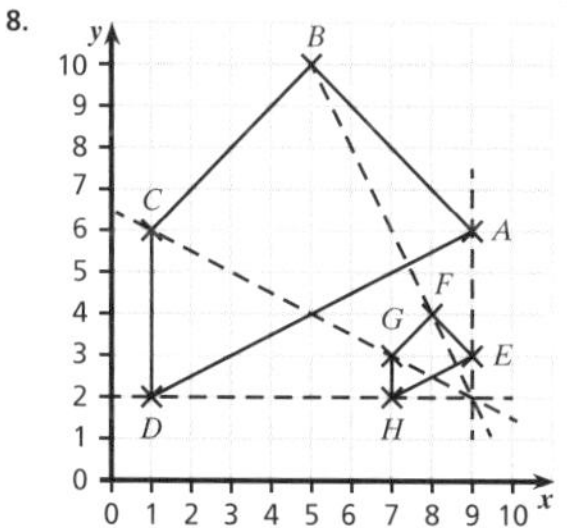


   a) Correct drawing of $ABCD$ **[1]**
   b) Correct drawing of $EFGH$ **[1]**
   c) Enlargement **[1]**; scale factor $\frac{1}{4}$ **[1]**; centre of enlargement (9, 2) **[1]**
9.

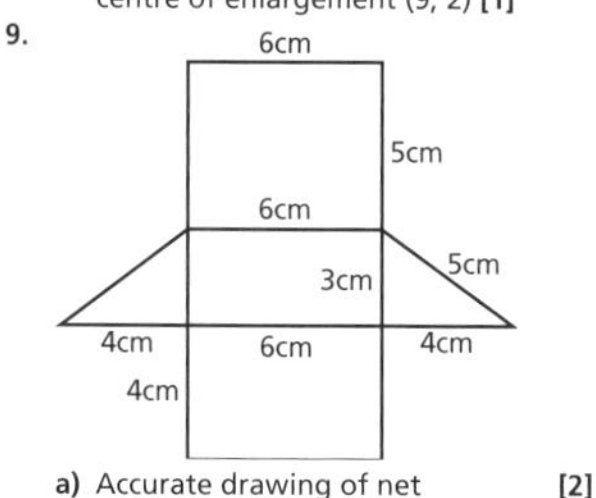


   a) Accurate drawing of net **[2]**
   b) 5 faces **[1]**
10. Sketch of 3D shape with matching front view **[1]**; side view **[1]**; and plan view **[1]**.

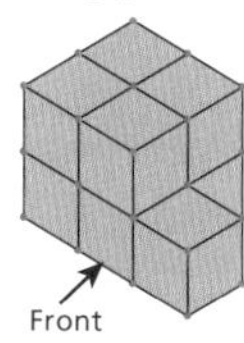


**Page 25**

1. a) Perimeter = $7 + (2 \times 1) + (2 \times 5) + 3 + (2 \times 2)$ (or equivalent) **[1]**; 26cm **[1]**
   b) Area = $(7 \times 1) + (5 \times 3)$ **[1]**; 22cm² **[1]**
2. $6 \times 10 \times x = 300$ **[1]**; $x = 5$cm **[1]**
3. Area of square = 36cm² **[1]**; Area of circles = $2 \times \pi \times 1.5^2 = 14.137...$ **[1]**; Shaded region = 36 – 14.137... **[1]**; = 21.9cm² (to 3 significant figures) **[1]**
4. a) Volume of large cylinder = $\frac{3}{4} \times 6000\pi = 4500\pi$ **[1]**; $4500\pi = \pi \times 15^2 \times h$ **[1]**; $h = 20$cm **[1]**
   b) Volume of small cylinder = $\frac{1}{4} \times 6000\pi = 1500\pi$ **[1]**; $1500\pi = \pi \times r^2 \times h = \pi \times r^3$ **[1]**; $r = 11.4$cm (to 3 significant figures) **[1]**
5. Volume of cone = $\frac{1}{3} \times \pi \times 3^2 \times 7 = 21\pi$ **[1]**; Volume of the hemisphere = $\frac{1}{2} \times \frac{4}{3} \times \pi \times 3^3 = 18\pi$ **[1]**; Volume of plastic needed = $21\pi + 18\pi = 39\pi$ cm³ **[1]**

## Pages 26–31 Revise Reading

**Page 27 Quick Test**

1. a) Angle $ABC$ = Angle $ADE$ (corresponding angles); Angle $ACB$ = Angle $AED$ (corresponding angles); Angle $DAE$ is common to both triangles; so triangles are similar (three matching angles).
   b) $\frac{5}{10} = \frac{BC}{8}$ so $BC$ = 4cm

**Page 29 Quick Test**

1. 13cm
2. $8^2 + 15^2 = 17^2$
3. $24^2 + 56^2 = C^2$, $C = 60.9$cm

**Page 31 Quick Test**

1. $\sin\theta = \frac{3.5}{8} = 0.4375$, $\theta = 25.9°$
2. $\cos 35° = \frac{x}{14}$, $x = 11$km

## Pages 32–35 Review Questions

**Page 32**

1. a) $X = (-5, 1)$ **[1]**; $Y = (-3, 5)$ **[1]**; $Z = (-1, 2)$ **[1]**
   b) $X = (1, -5)$ **[1]**; $Y = (5, -3)$ **[1]**; $Z = (2, -1)$ **[1]**
2. Lengths of rectangle D: 3 × 3 = 9cm and 5 × 3 = 15cm **[1]**; Area of rectangle C = 3 × 5 = 15cm², area of rectangle D = 9 × 15 = 135cm² **[1]**; Ratio = 15 : 135 = 1 : 9 **[1]**
3. Volume of cuboid C = 3 × 4 × 5 = 60cm³ **[1]**; Volume of cuboid D = 9 × 12 × 15 = 1620cm³ **[1]**; Ratio = 60 : 1620 = 1 : 27 **[1]**
4. $A' = (3, 2)$ **[1]**; $C' = (6, 6)$ **[1]**; $B' = (5, 3)$ **[1]**; $D' = (4, 5)$ **[1]**
5. a) A vertical line **[1]**
   b) An arc of a circle **[1]**
   c) A horizontal straight line **[1]**
   d) An arc of a circle **[1]**
6. A square 4cm × 4cm **[1]**
7.

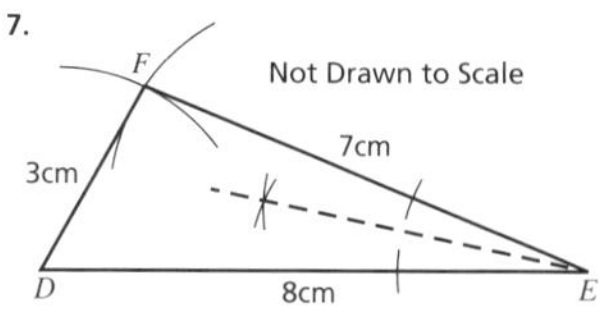


   a) Correct construction of triangle (see above) **[2]**
   b) Angle $FDE$ = 60° (+ or – 2°) **[1]**
   c) Correct angle bisector of $FDE$ **[2]**
8. a) **[2]**
   b) **[2]**
   c) **[2]**

9. Plan Side
   a) Correct plan view **[2]**
   b) Correct side elevation **[2]**
10. a) CHOKE **[1]**
   b) Any appropriate word with a horizontal line of symmetry, e.g. BED, HIKE, BOX, BID **[1]**

**Page 34**

1. a) $\frac{1}{2} \times 6 \times 8 \times 9$ **[1]**; 216cm³ **[1]**
   b) $\sqrt[3]{216} = 6$ **[1]**; $6 \times 12$ **[1]**; = 72cm **[1]**
2. $\pi r^2 = 2\pi r$ **[1]**; $r = 2$ **[1]**
3. $\frac{1}{2}(2x + x) \times 3x \times 20 = 900$ **[1]**; $9x^2 = 90$ **[1]**; $x^2 = 10$ **[1]**; $x = \sqrt{10}$ or 3.16cm **[1]**
4. $75 = 4 \times \pi \times r^2$ **[1]**; $r^2 = \frac{75}{4\pi}$ **[1]**; $r = 2.44$cm (to 3 significant figures) **[1]**
5. $\frac{1}{2} \times x \times 6 = 7.5$ **[1]**; $6x = 15$ **[1]**; $x = 2.5$ **[1]**
6. 5 × 3.5 = 17.5m² **[1]**; 0.5 × 5 × 1 = 2.5m² **[1]**; 17.5 + 2.5 = 20m² **[1]**; 2 tins **[1]**

**Page 35**

1. a) $\frac{24}{6}$ **[1]**; 4cm **[1]**
   b) 2 × 6 + 2 × 4 **[1]**; = 20cm **[1]**
2. 6 × 4 = 24 **[1]**; $\frac{24}{3}$ **[1]**; 8cm **[1]**

# Answers

## Pages 36–37 Practice Questions

**Page 36**

1. Angle $ACB$ = Angle $BDC$ = 90° **[1]**; Angle $ABC$ = Angle $DBC$ (the angle is common to both triangles) **[1]**; Angle $BAC$ = Angle $BCD$ (180° – Angle $B$ – 90°), so the triangles are similar (three matching angles) **[1]**
2. a) 5cm **[1]**
   b) 20cm **[2]**
3. a) A and B **[1]**; AAS OR SAS **[1]**
   b) B and D

**Page 37**

1. $7.6^2 + 6.7^2 = y^2$ **[1]**; $57.76 + 44.89 = y^2$ **[1]**; $y = \sqrt{(102.65)} = 10.13$km **[1]**
2. $4^2 - 2^2 = f^2$ **[1]**; $16 - 4 = f^2$ **[1]**; $\sqrt{12} = f$, fence height = 3.46m **[1]**; No, because the fence height is 3.46m and Fang can only jump 3m. **[1]**
3. $13^2 - 5^2 = BD^2$ **[1]**; $169 - 25 = BD^2$, $BD = \sqrt{144}$ **[1]**; $BD$ = 12cm **[1]**
4. $10^2 + 6^2 = b^2$ **[1]**; $100 + 36 = b^2$ **[1]**; $b = \sqrt{136} = 11.7$m (to 3 significant figures) **[1]**
5. $1.5^2 + 2^2 = 6.25$ **[1]**; $\sqrt{6.25} = 2.5$ **[1]**; The triangle is right-angled (Pythagoras' Theorem) **[1]**.
6. $3^2 + 3^2 = d^2$ **[1]**; $d = \sqrt{18} = 4.24$cm **[1]**
7. $\tan\theta = \frac{5}{3} = 1.6667$ **[1]**; $\theta = 59°$ **[1]**
8. a) $5^2 + 8^2 = CA^2$, $89 = CA^2$ **[1]**; $CA = \sqrt{89}$ = 9.4km **[1]**
   b) $\sin CAB = \frac{8}{9.4} = 0.851$ **[1]**; $CAB = 58°$ **[1]** OR $\tan CAB = \frac{8}{5} = 1.6$ **[1]**; $CAB = 58°$ **[1]**

Making a sketch will show that you are dealing with a right-angled triangle.

9. a) $\sin 32° = \frac{GH}{9}$ **[1]**;
   $GH = \sin 32° \times 9 = 4.77$cm **[1]**
   b) $\tan HGJ = \frac{6}{4.77}$ **[1]**;
   $\tan HGJ = 1.2579$, 51.5° **[1]**

## Pages 38–41 Revise Questions

**Page 39 Quick Test**

1. Angle $XOY$ = 100°
2. Circle of radius 4cm

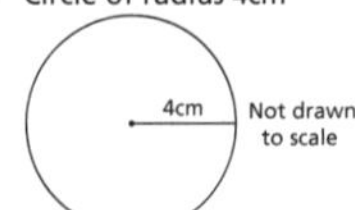


**Page 41 Quick Test**

1. a)

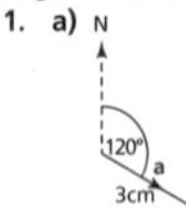


b) N, 4cm, 40°, b

c) N, –a

d) N, 2a

e) N, a + b, a, b

## Pages 42–43 Review Questions

**Page 42**

1. 56° **[1]**; 64° **[1]**; 60° **[1]**
2. Shadow is $\frac{34}{8.5}$ times bigger = 4 times bigger **[1]**; height of tree = 5 × 4 = 20m **[1]**
3. Similar figures are identical in shape but can differ in size **[1]**; congruent figures are identical in shape and size. **[1]**
4. AAS **[1]**
5. $\frac{38}{9.5} = \frac{\text{height of lamp post}}{4}$ **[1]**; lamp post = 16m **[1]**
6. Perimeter of A = 38cm **[1]**; perimeter of B = 114cm, so scale factor = 3 **[1]**

**Page 43**

1. $10^2 + 5^2 = d^2$ **[1]**; $d = \sqrt{125}$, $d$ = 11.2cm **[1]**
2. $x^2 + x^2 = 12^2$ **[1]**; $2x^2 = 144$, $x^2 = 72$ **[1]**; $x = \sqrt{72} = 8.49$cm **[1]**
3. a) Right-angled triangle or an isosceles triangle **[1]**
   b) $PR$ = 5 units **[1]**
   c) $PQ$ = 5 units **[1]**
   d) $5^2 + 5^2 = QR^2$ **[1]**; $QR^2 = 50$, $QR = \sqrt{50}$ **[1]**; $QR$ = 7.07 units **[1]**
   e) Area = $\frac{1}{2}$(base × height) **[1]**; $\frac{1}{2} \times 5 \times 5 = 12.5$ square units **[1]**
4. Sean was correct **[1]**; $9^2 = 81$, $40^2 = 1600$, $41^2 = 1681$ so $9^2 + 40^2 = 41^2$ **[1]** Pythagoras' Theorem works, so triangle is right-angled. **[1]**
5. $\tan\theta = \frac{\text{opp}}{\text{adj}}$, $\tan 63.15° = \frac{x}{80}$ **[1]**; $x = 80 \times \tan 63.15°$ **[1]**; $x$ = 158m **[1]**
6. $\tan\theta = \frac{3}{5}$ **[1]**; $\tan\theta = 0.6$, $\theta = 30.964°$ **[1]**
7. sin 30° + sin 60° = 0.5 + 0.8660 = 1.366 **[1]**; cos 30° + cos 60° = 0.8660 + 0.5 = 1.366 **[1]**; Both answers are the same – Chevaun is incorrect **[1]**
8. a) $\sin 45° = \frac{x}{175}$ **[1]**;
   $x = 175 \times \sin 45° = 123.7$km **[1]**;
   $x$ = 124km **[1]**
   b) 100km in 30 minutes, 50km in 15 minutes, 25km in 7.5 minutes **[1]**; 175km in 52.5 minutes **[1]**

## Page 44 Practice Questions

**Page 44**

1. a) 90° **[1]**
   b) 45° **[1]**
   c) 40° **[1]**
   d) 60° **[1]**
   e) 100° **[1]**
   f) 90° **[1]**

1. a) Right-angled triangle **[2]**
   b) (1, –2) **[1]**; (5, –2) **[1]**; (5, –4) **[1]**
   c) (–6, 5) **[1]**; (–2, 5) **[1]**; (–2, 3) **[1]**

## Page 45 Review Questions

**Page 45**

1. a) Arc **[1]**
   b) Radius **[1]**
   c) Circumference **[1]**
   d) Diameter **[1]**
   e) Chord **[1]**
   f) Tangent **[1]**
   g) $\pi$ **[1]**
   h) $A = \pi r^2$ **[1]**
   i) $C = \pi d$ **OR** $2\pi r$ **[1]**
2. **8a + 6b** **[1]**
3. **Correctly translated shape [1]**

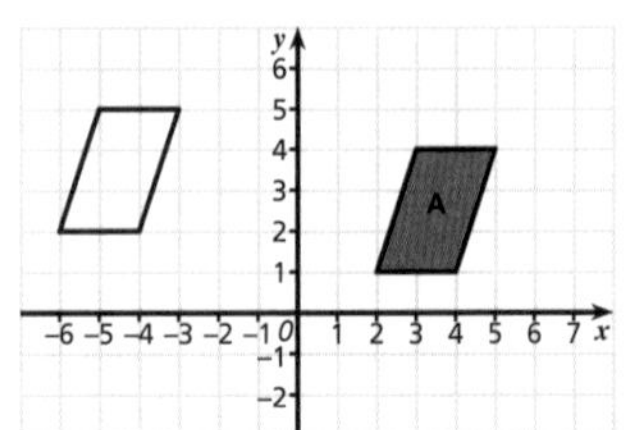